Practical interventional radiology of the hepatobiliary system and gastrointestinal tract

Practical interventional radiology of the hepatobiliary system and gastrointestinal tract

Edited by

Andy Adam, MRCP, FRCR
Department of Radiology, Guy's Hospital, London
and
Robert N. Gibson, FRACR, DDU
Department of Radiology, Royal Melbourne Hospital, Melbourne

Edward Arnold
A member of the Hodder Headline Group
LONDON BOSTON MELBOURNE AUCKLAND

© 1994 Edward Arnold

First published in Great Britain 1994

Distributed in the United States by Little, Brown and Company,
34 Beacon Street, Boston, MA 02108

British Library Cataloguing in Publication Data
Available on request
ISBN 0-340-55166-6

Photoset in Linotron 202 Bembo by Rowland Phototypesetting
Limited, Bury St Edmunds, Suffolk.
Printed and bound in Great Britain for Edward Arnold, a division of
Hodder Headline PLC, Mill Road, Dunton Green, Sevenoaks, Kent
TN13 2YA by Butler & Tanner Limited, Frome, Somerset

Preface

In the liver and biliary tract, perhaps more than in any other organ system, diseases are best managed by a team rather than individual specialists in isolation. Cooperation between surgeons, interventional radiologists and gastrointestinal endoscopists is frequently necessary in solving complex problems. The hepatobiliary interventional radiologist must have a good understanding of the surgical and endoscopic techniques used in the liver and biliary tract, and he needs to have a detailed knowledge of the relevant anatomy and must be familiar with the available interventional radiological equipment. The latter requirement is a very important one: in recent years companies manufacturing catheters and related equipment have realized that the requirements of the nonvascular interventional radiologist differ significantly from those of the vascular interventionalist. This recognition has lead to the development of purpose-built instruments which have made procedures safer and have expanded the possibilities of intervention in this field.

This volume does not aim to provide a detailed scientific analysis of the results of the various procedures described. We have chosen instead to concentrate on the practical considerations underlying the various techniques and have tried to provide some useful practical guidelines. All the authors are very experienced interventional radiologists or endoscopists currently active in their fields. The procedures and equipment described are those being used by the authors; we have omitted detailed descriptions of outdated methods and techniques.

We hope that this volume will prove useful to interventional radiologists who would like to increase their knowledge of procedures in the hepatobiliary system and gastrointestinal tract.

The original suggestion to write this book came from Professor Graham Whitehouse, Professor of Radiology at the University of Liverpool. We are very grateful to him for outlining the concept and providing the initial inspiration.

A. Adam
R. N. Gibson
1993

Contents

Preface   v

Contributors   ix

Section 1 The biliary tree and gall bladder

1   Removal of retained biliary calculi   3
     Eugene Y. Yeung

2   Percutaneous transhepatic
     cholangiography and biliary
     drainage   13
     Robert N. Gibson

3   Biliary endoprosthesis   32
     Andy Adam

4   Interventional radiology for benign
     biliary strictures   52
     Robert N. Gibson and Andy Adam

5   Percutaneous gall bladder
     procedures   67
     Eugene Y. Yeung

6   Endoscopic and combined
     management of biliary strictures   81
     *Antony G. Speer and
     Robert N. Gibson*

Section 2 Biopsy and abscess drainage

7   Computed tomographic-guided and
     ultrasound-guided intra-abdominal
     biopsy and abscess drainage   99
     Eugene Y. Yeung

8   Liver biopsy in patients with
     abnormal coagulation: alternatives
     to the transjugular approach   118
     Andy Adam

**Section 3 Oesophageal and gastrointestinal
             intervention**

9   Oesophageal strictures   128
     *Steven G. Meranze and
     Gordon K. McLean*

10   Gastrointestinal strictures and
     fistulae   140
     Dana Burke and Gordon K. McLean

Section 4 Angiography

11   Embolization of the liver   175
     *Götz Richter and
     Gunter W. Kauffmann*

12   Vascular stents in liver disease   189
     Götz Richter and J. C. Palmaz

Index   207

Contributors

Andy Adam, MRCP, FRCR,
Department of Radiology, Guy's Hospital, London, UK

Dana Burke, MD,
St Luke's Hospital, Bethlehem, Pennsylvania, USA

Robert Gibson, FRACR, DDU,
Department of Radiology, Royal Melbourne Hospital, Melbourne, Australia

Gunter W. Kauffman,
Oberarzt der Abteilung Radioagnostik (Chirurgie), Uniklinik Heidelberg, Heidelberg, Germany

Gordon K. McLean, MD,
Department of Angiography and Interventional Radiology, West Penn Hospital, Pittsburgh, Pennsylvania, USA

Steven G. Meranze, MD,
Department of Radiology, Vanderbilt University, Nashville, Tennessee, USA

J. C. Palmaz,
Department of Radiology, University of Texas Health Science Center at San Antonio, USA

Götz Richter,
Oberarzt der Abteilung Radioagnostik (Chirurgie), Uniklinik Heidelberg, Heidelberg, Germany

Antony G. Speer, FRACP,
Department of Gastroenterology, Royal Melbourne Hospital, Melbourne, Australia

Eugene Y. Yeung, FRCPC,
Department of Radiology, The Toronto Hospital, Toronto, Canada

1

THE BILIARY TREE AND GALL BLADDER

Removal of retained biliary calculi

Eugene Y. Yeung

Introduction   4

Extraction through a T-tube tract   4

Adjuvant and alternative techniques   9

References   11

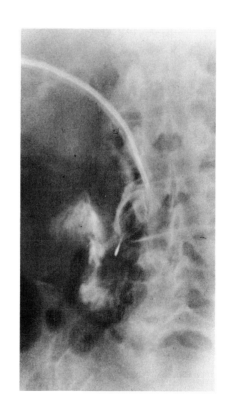

INTRODUCTION

It has been estimated that approximately 5% of postoperative T-tube cholangiograms will show retained calculi (Glenn, 1974). If retained stones are detected, percutaneous removal through the T-tube tract, as popularized by Mazzariello (1978) and Burhenne (1980), is the treatment of choice. If no T-tube is present, stones may be removed through an endoscopic or transhepatic approach.

EXTRACTION THROUGH A T-TUBE TRACT

Patient preparation

Extraction through a T-tube tract should not be performed for at least 4 weeks after surgery to allow the T-tube tract to mature. Thus, the patient is usually discharged home and undergoes stone extraction as an outpatient, often with only minimal analgesia. An intravenous (i.v.) line should be established so that i.v. sedation and analgesia may be given if required. Antibiotic prophylaxis is used routinely by some radiologists (e.g. 1 g piperacillin i.v. just prior to the procedure).

Technique

INITIAL CHOLANGIOGRAM

A cholangiogram is performed to assess the number, size and position of bile duct stones. Care should be taken not to introduce air bubbles into the biliary tree as they may mimic stones on cholangiography. Dilute contrast medium should be used so that stones are not obscured (e.g. 30% Hypaque, Sterling Winthrop, New York, USA). If the bile ducts are markedly dilated, more diluted contrast medium should be injected; however, overdistension should be avoided. When the correct volume of contrast medium is injected, the entire intrahepatic and extrahepatic biliary tree should be visible. A variety of views of the biliary tree should be obtained (with the patient supine, turned 45° to the left side and completely left side down) to allow better appreciation of the presence and location of any stones.

STONE EXTRACTION

The procedure is usually performed from the patient's right side. Most surgeons now use T-tubes at least as large as 14 Fr, and bring the tube out in a straight path close to the right costal margin in case percutaneous extraction is necessary. If the tract has been brought out anteriorly in the epigastrium, stone extraction should be performed with the patient placed at an oblique angle to the right so the tract is not foreshortened on fluoroscopy.

A slightly curved standard 0.038 inch guidewire is introduced into the T-tube (Figs 1.1 and 1.2). It is advanced either upwards into the intrahepatic ducts or downwards into the common bile duct (CBD). Slight traction on the T-tube facilitates passage of the guidewire tip past the T-junction of the tube. With further steady traction the T-tube is removed over the wire.

Depending on the diameter of the T-tube and the size of the retained stones, the operator should decide whether the tract requires dilation. Generally, 5–10 mm stones will pass through an 18 Fr tract. Larger stones may require either tract enlargement or fragmentation. The adequacy of the tract size can usually be estimated by comparing the tract and stone size on the initial cholangiogram.

Tract dilation is performed using serial dilators, and a peel-away sheath may be inserted with the final dilation to maintain the tract. Steerable catheters (Meditech, Watertown, MA) are used for manipulation within the biliary tree. The steerable catheter is inserted over the guidewire and manoeuvred into the appropriate part of the duct. The tip of the catheter should be negotiated just past the stone providing there is sufficient space.

The stone basket sheath is inserted along the guidewire, which is then removed. Contrast medium should be gently injected again at this point to delineate the stone. Brisk injection should be avoided as it may flush the stone into a less favourable position for extraction. Air bubbles that hinder visualization may be removed by aspiration through the catheter. The stone basket is then inserted to the end of the sheath. An exchange technique is used to withdraw the steerable catheter and the sheath, allowing the basket to open over the stone. The stone is engaged by rotating and moving the basket to and fro. If the CBD is grossly dilated and the stone floats freely, the technique illustrated in Fig. 1.3 should be used.

Once the stone is in the basket, the steerable catheter, basket and sheath are withdrawn as one unit, and the stone is removed through the tract. Withdrawal should be steady and along the line of

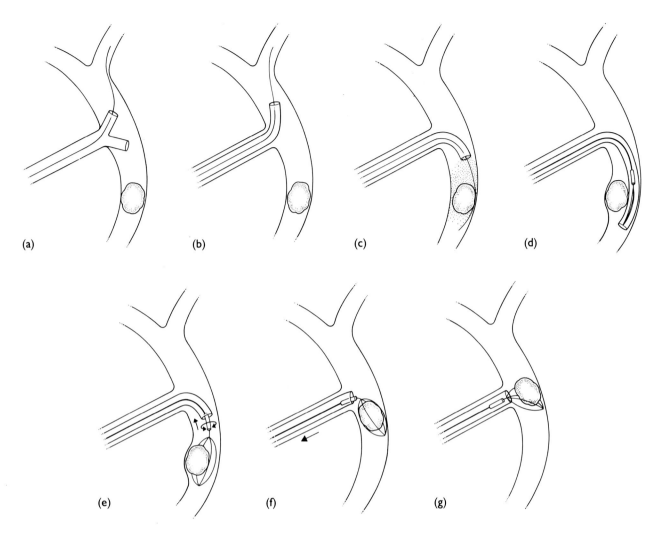

(a)   (b)   (c)   (d)

(e)   (f)   (g)

Fig. 1.1 *Standard technique of T-tube stone extraction*
(a) After a cholangiogram locates the retained stone(s), a slightly curved tipped guidewire is inserted through the T-tube. Slight traction on the T-tube facilitates passage of the wire through one limb. (b) The T-tube is removed and a steerable catheter inserted. (c) The steerable catheter is manipulated into the vicinity of the stone, contrast medium is injected again to maintain visualization and the guidewire is negotiated past the stone. (d) The catheter is advanced past the stone. (e) The basket sheath is inserted over the guidewire, which is then removed, and the basket inserted to lie adjacent to the stone. (f) Once the stone is engaged, the steerable catheter is slowly withdrawn as one unit. (g) Care should be taken to ensure that the steerable catheter does not override the basket causing it to close prematurely. This often has the effect of squeezing the stone out of the basket, especially when negotiating the bend between the bile duct and T-tube tract.

the tract to prevent fragmentation or dislodgement. It is not usually necessary to advance the catheter and sheath to grip the stone in the basket; in fact, this may precipitate dislodgement.

The steerable catheter is reinserted along the T-tube tract into the bile duct and further stones are removed in a similar fashion. Insertion over a guidewire is not usually necessary because the T-tube tract should be well formed. Once all the stones have been removed, a cholangiogram is performed. If no filling defects are observed in the entire biliary tree and contrast medium is draining freely into the duodenum, the catheter is removed and the procedure terminated. Frequently, however, debris and blood clot are present. In such a case, a

suitably sized red rubber catheter (usually 12–18 Fr) should be left in the biliary tree so that another cholangiogram may be performed in 2 to 4 days to check for residual stones.

CHOICE OF BASKET

A stone basket is selected according to the size of the stone and the diameter of the bile duct (Fig. 1.4). In general, the basket should occupy the width of the bile duct when open. Baskets also vary in the number and type of wires and wire configuration. Larger stones are usually trapped with four-wire helical baskets and smaller ones with eight-wire baskets. It is useful to check the shape of the basket

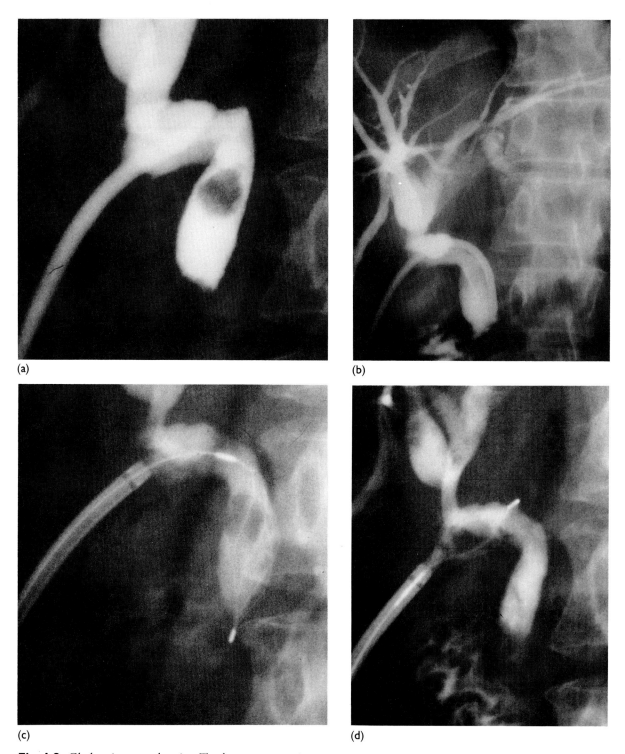

Fig. 1.2 *Cholangiograms showing T-tube stone extraction*
(a) A stone is visible in the common bile duct. (b) A steerable catheter is manipulated past the stone. (c) A helical stone basket is opened over the stone. (d) The stone basket and steerable catheter are removed as one unit.

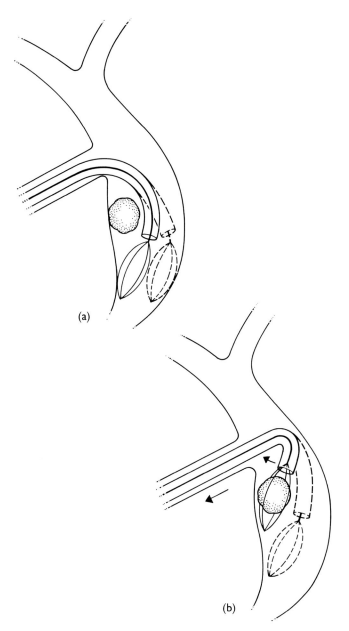

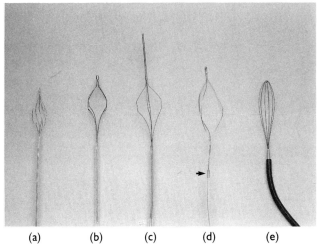

Fig. 1.4 *Examples of stone baskets available for use.*
(a) Eight-pronged helical. (b) Four-pronged shape-reforming multifilament. (c) Four-pronged multifilament with filiform tip for use through the ampulla. (d) Three-pronged helical with steel ringed tip for stone fragmentation (arrow). (e) Nitinol shape-reforming basket.

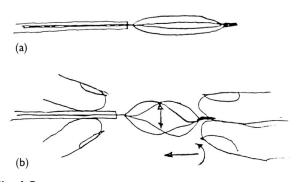

Fig. 1.5
(a) The stone basket has lost its helical shape after usage. (b) The basket shape is reformed by grasping the tip with the index finger and thumb and exerting a forward and twisting motion.

Fig. 1.3 *When the common bile duct is grossly dilated and the stone floats freely, it is captured by opening the basket distal to it*
(a) The steerable catheter tip is then curved in the appropriate direction. (b) While that angle is maintained, the catheter and basket are withdrawn over the stone as one unit.

and correct any misaligned wires before insertion (Fig. 1.5).

Aftercare

Minimal aftercare is necessary. Broad-spectrum antibiotics may be given orally for 3 to 5 days if the stone extraction is relatively traumatic. The tract usually closes in several days. The patient should be asked to report any fever or abdominal pain.

Special considerations

IMPACTED STONES

A stone impacted proximal to the ampulla in the CBD may be difficult to remove. Disimpaction can be attempted by negotiating a wire past the stone into the duodenum. A standard straight-tipped 0.038 inch wire or glide wire (Terumo, Meditech) should be used. An occlusion balloon (Fig. 1.6) is inserted over the guide wire and then inflated distal to it. On retraction, the stone is pulled into the wider

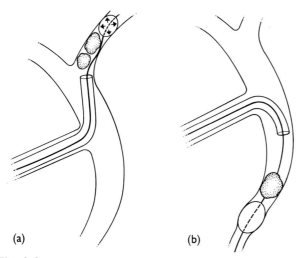

Fig. 1.6
(a) Occlusion balloons may be used to manipulate an intrahepatic stone into a more favourable position. (b) They can also be used to dislodge impacted preampullary stones.

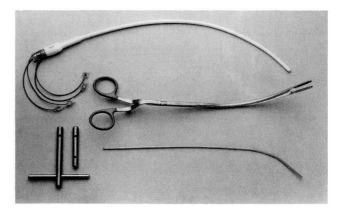

Fig. 1.7
Instruments used for stone extraction. Meditech steerable catheters (top) are available in sizes from 8.5 to 13 Fr. Generally, 13 Fr is most useful for T-tube stone extraction. The catheter is 'steered' by traction on one of the four wires located at its hub. Mazzariello forceps (middle) are designed for use through long tracts. Different sizes and angled shafts are also available. A simple mechanical lithotripter (bottom) (Ho *et al.*, 1987) consists of a slightly curved rigid metallic cannula, a crossbar and a T-handle (manufactured by Homach, Toronto, ON; distributed by Beejay Medical, Oakville, ON, Canada).

proximal bile duct allowing basket removal. The use of occlusion balloons may not be successful if the stone is tightly impacted.

Alternatively, a basket may be used to engage the impacted stone, although the basket may not open sufficiently to capture it because of lack of space. Also, an open basket must be manoeuvred cautiously through the ampulla because the wires of the basket may snare folds of mucosa, risking damage and precipitating pancreatitis. If a guidewire cannot be negotiated past the stone, choledochoscopy and intracorporeal fragmentation are recommended.

INTRAHEPATIC STONES

A stone in the main right or left hepatic duct may also be removed by negotiating a guide wire past it and using a stone basket. If stones are in second- or third-order intrahepatic ducts, it may be more expedient to manipulate them into the CBD with an occlusion balloon (Fig. 1.6).

FRAGMENTATION

Large stones require fragmentation before extraction. This can be achieved by engaging the stone within a basket and mechanically fragmenting it (mechanical lithotripsy). Special stone baskets with steel ringed tips (Wilson Cook, Winston-Salem, NC) are available (Fig. 1.4). Five minutes of traction using these baskets is usually sufficient to fragment the stone. Alternatively, the stone may be engaged in a regular stone basket and then fragmented by

threading a specially designed rigid, metallic cannula over the basket wire (Fig. 1.7). Gradual retraction of the basket against the cannula fragments the stone (Ho *et al.*, 1987) (Figs 1.8 and 1.9). Flexible metallic devices that work in a similar manner are also available, but they are more cumbersome (Soehendra mechanical lithotripter, Cook, Bloomington, IN).

Results and complications

Percutaneous stone extraction through a T-tube tract is generally straightforward. The overall success rate is about 95% (Mazzariello, 1978; Burhenne, 1980). Failures are most often due to an inability to capture or fragment stones in awkward positions (e.g. cystic duct remnant).

The complication rate has been reported to be about 4% (Burhenne, 1980). Most complications are minor and include cholangitis and T-tube tract disruption with potential peritoneal spillage. Loss of stone fragments within the tract may also occur but this is not usually serious. Pancreatitis induced by manipulation at the distal CBD is probably the most serious complication. Fortunately, it is rare, although fatalities have been reported (Mazzariello, 1978).

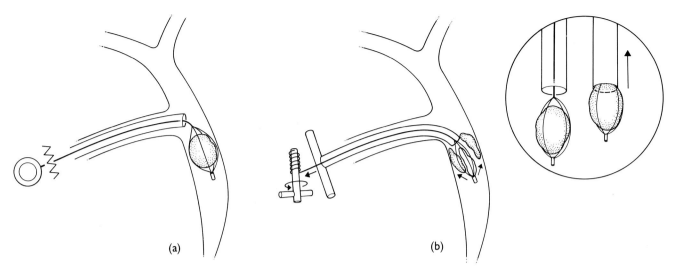

Fig. 1.8 *Technique of mechanical lithotripsy (Ho et al., 1987)*
(a) Once the stone has been trapped in the basket, the external part of the basket wire and sheath are cut. (b) The metal cannula is threaded carefully over the basket sheath and wire to the stone. The crossbar is threaded onto the cut wire, which is then attached to the T-handle. Rotating the T-handle acts as a simple winch progressively tightening the basket, which eventually slices through the stone, fragmenting it (inset).

ADJUVANT AND ALTERNATIVE TECHNIQUES

(see also Chapter 4)

Waiting and irrigation

Many stones in awkward positions not amenable to extraction will eventually migrate to more favourable sites. Repeating the procedure in 3 to 5 days often allows successful extraction.

When there are numerous stones and debris filling multiple ducts, infusion of saline (e.g. 500 ml normal saline) through an appropriately placed catheter (e.g. in an intrahepatic bile duct) over 24 hours may help to reduce the number of stones. However, this should be done only if the CBD is not obstructed.

Pulsed jet saline irrigation is also effective in dislodging impacted or adherent stones. In this technique, a peel-away sheath is inserted into the T-tube tract. A biliary manipulation catheter (see Fig. 2.14, p. 24) is advanced through the sheath and positioned adjacent to the stones. The catheter is connected by tubing to a household dental hygiene unit that provides the pulsed jet source (e.g. Water-Pik, Teledyne, Los Angeles, CA). The jet is activated and the catheter is manipulated using to-and-fro and twisting motions. Irrigation fluid, disimpacted stone fragments and debris flow out around the catheter through the peel-away sheath.

Flushing stones into the duodenum by forceful injection through the T-tube after administration of Buscopan (Boehringer, Ingelheim, Germany) to relax the sphincter of Oddi has also been tried (Catt et al., 1974). This procedure is not recommended because the unphysiologically high biliary pressures it produces frequently cause rigors and sepsis.

Forceps

Forceps cannot generally be used for T-tube stone extraction because of the length of the tract and angulation of the bile ducts. Mazzariello forceps (American Hospital Supplies, Chicago, IL), however, are designed specifically for use through long tracts and can be used if basket removal is not successful or for retrieval of stones dislodged in the tract (Fig. 1.7).

Choledochoscopy and intracorporeal lithotripsy

Flexible choledochoscopy through the T-tube tract is becoming more popular. The 16 Fr choledochoscope (Olympus, Tokyo, Japan) is easily used through an 18 Fr peel-away sheath (Cook). It allows excellent visualization of the extrahepatic biliary tree and first- and second-order intrahepatic ducts. Blood clot and debris are also readily distinguished

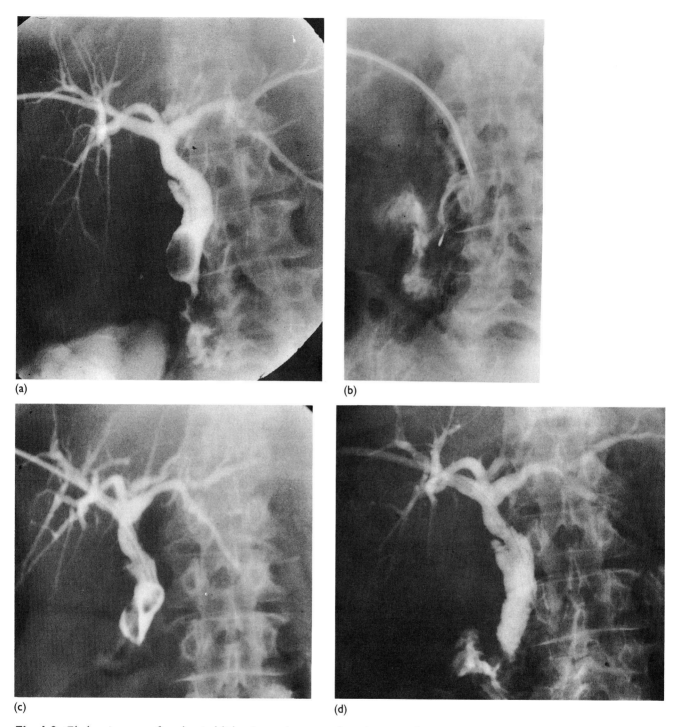

Fig. 1.9 *Cholangiograms of mechanical lithotripsy using a transhepatic approach*
(a) The view through a right percutaneous drainage catheter shows a large impacted common bile duct stone (which could not be removed with endoscopic retrograde cholangio pancreatography. (b) The stone has been trapped in a basket and the rigid metallic cannula of the mechanical lithotripsy device inserted to the stone. (c) After fragmentation. (d) The final cholangiogram shows clearance of fragments from the common bile duct. Reproduced from Ho *et al.* (1987) with permission.

from stones, which may often be difficult using fluoroscopy alone. Furthermore, fragmentation of large or impacted stones is easily achieved through the use of intracorporeal lithotripsy (e.g. electro-hydraulic lithotripsy wire or laser lithotripsy fibreoptic filament, both inserted through the choledochoscope working channel).

Chemical dissolution

Stone dissolution with methyl *tert*-butyl ether (MTBE), mono-octanoin and other solvents has been reported. These agents have been infused into the biliary tree through T-tubes and nasobiliary or percutaneous biliary catheters. Generally, these agents have been unsatisfactory for several reasons. Firstly, only cholesterol-containing stones are amenable to dissolution [25% are pigment stones (Bills and Lewis, 1975)]. Secondly, successful dissolution may take a prolonged period (at least hours, often days). Thirdly, these agents have potentially significant side effects and complications.

Mono-octanoin requires infusion over prolonged periods (often more than 1 week) for an adequate effect, and its use has also been associated with diarrhoea, vomiting and abdominal cramps. Furthermore, it was reported to be successful in only 48% of 475 patients treated (Schmack, 1983).

Methyl *tert*-butyl ether is a more potent cholesterol solvent and early human studies have been encouraging. Dissolution is much faster than with mono-octanoin, but still requires many hours of infusion. However, MTBE is flammable and must be used cautiously. It is also toxic and can cause duodenal ulceration and systemic side effects if absorbed through the bowel. Because of these side effects, MTBE should be used only if an occlusion balloon is inflated in the bile duct to prevent spillage into the duodenum.

Thus, chemical dissolution has not been widely adopted and is unlikely to become so. Choledochoscopy and direct fragmentation are more efficient and less time consuming and have fewer potential complications.

Endoscopic or transhepatic removal

The preferred approach to removal of CBD stones in patients who do not have a T-tube in place is by endoscopic sphincterotomy and basket removal. This is effective in 90% of patients, but if stones are more than 1.5 cm in diameter or located in the intrahepatic ducts, alternative techniques are often necessary. Alternatives include chemical dissolution through a nasobiliary tube, extracorporeal shock-wave lithotripsy (ESWL), percutaneous transhepatic removal and, ultimately, surgery. Chemical dissolution and its potential side effects have been discussed above. Extracorporeal shockwave lithotripsy of bile duct stones requires contrast medium opacification of the CBD through a nasobiliary tube for targeting. Endoscopic sphincterotomy is also necessary to allow the stone fragments to pass. However, additional endoscopic removal of fragments may still be required. Thus, ESWL is unlikely to have a major impact on the treatment of bile duct stones.

Percutaneous transhepatic removal of bile duct calculi has recently gained attention as a possible approach (Gandini *et al.*, 1990; Stokes and Clouse, 1990; Ho *et al.*, 1992). Percutaneous biliary drainage is performed through the right or left intrahepatic bile ducts. The duct chosen depends on the position of the stones. In general, a relatively horizontal approach to a right main duct allows access to the CBD. Although puncture of the central bile ducts should be avoided whenever possible to reduce the risk of hepatic artery or portal vein injury, in some cases, puncture of the right main or common hepatic duct is necessary. Fortunately, the common hepatic duct is usually anterior and lateral to the portal vein and hepatic artery at the liver hilum so it is entered first. Ultrasound guidance for the initial puncture is also recommended. Once drainage has been performed, the tract is allowed to mature for 1 to 2 days. With this access route, various methods can be used to fragment, flush and dissolve the stones (e.g. basket removal, choledochoscopy and electro-hydraulic lithotripsy, chemical dissolution). Stone fragments are usually removed by antegrade passage through the ampulla. A prior endoscopic or trans-hepatic sphincterotomy or balloon dilation of the ampulla is usually required.

I thank Ms Edvige Coretti for her expert secretarial assistance in the preparation of this chapter.

REFERENCES

BURHENNE JH (1980). Percutaneous extraction of retained biliary tract stones. *American Journal of Roentgenology* 134: 889–98.

BILLS PM, LEWIS D (1975). A structural study of gallstones. *Gut* 16: 630–7.

CATT PB, HOGG DF, CLUNIE GJ, HARDIE IR (1974). Retained biliary calculi: Removal by a simple non-operative technique. *Annals of Surgery* 180: 247–51.

GANDINI G, RECCHIA S, FERRARIS A, FRONDA GR (1990).

Percutaneous removal of biliary stones. *Cardiovascular and Interventional Radiology* **13**: 245–51.

GLENN F (1974). Retained calculi within the biliary ductal system. *Annals of Surgery* **179**: 528–39.

HO C-S, YEE AC, MCLOUGHLIN MJ (1987). Biliary lithotripsy with a mechanical lithotripter. *Radiology* **165**: 791–3.

HO CS, YEUNG EY, THURSTON W, FINNEGAN P (1992). Percutaneous transhepatic management of biliary calculi: Review of 67 patients. Presented at the Annual Meeting of American Roentgen Ray Society.

MAZZARIELLO RM (1978). A fourteen-year experience with nonoperative instrument extraction of retained bile duct stones. *World Journal of Surgery* **2**: 447–55.

SCHMACK B (1983). Dissolution of bile duct stones. *Endoscopy* **15** (Suppl 1): 186–90.

STOKES KR, CLOUSE ME (1990). Biliary duct stones: Percutaneous transhepatic removal. *Cardiovascular and Interventional Radiology* **13**: 240–4.

CHAPTER 2

Percutaneous transhepatic cholangiography and biliary drainage

Robert N. Gibson

Percutaneous transhepatic
cholangiography   14

Percutaneous biliary drainage   18

References   31

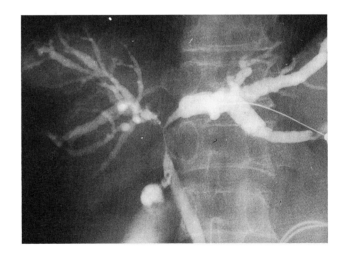

Percutaneous transhepatic cholangiography (PTC) and endoscopic retrograde cholangiography (ERC) are the most frequently used methods of direct cholangiography and provide the most complete and detailed radiographic demonstration of the bile ducts. In many centres ERC has substantially replaced PTC for diagnostic purposes and endoscopic stenting has also replaced percutaneous biliary drainage (PBD) for many patients. Percutaneous transhepatic cholangiography does, however, remain an important and integral part of radiological evaluation of the obstructed biliary system and PBD plays an important role in its treatment.

PERCUTANEOUS TRANSHEPATIC CHOLANGIOGRAPHY

Indications

Direct cholangiography is indicated in the presence of bile duct obstruction which has been demonstrated by ultrasound or computed tomography (CT), where the information provided by those studies is insufficient for diagnostic purposes or for planning treatment. The choice between PTC and ERC will be determined by local availability and expertise and by factors listed in Table 2.1.

In general ERC is preferable if there is no intra-hepatic duct dilation or if *low* common bile duct obstruction is suspected. For suspected *high* bile duct obstruction ERC may be used initially but it should be recognized that this often provides incomplete information about the bile ducts above a high obstruction and PTC may then be necessary for diagnostic purposes and treatment planning. Percutaneous transhepatic cholangiography is the preferred method if segmental intrahepatic duct obstruction is suspected or if the second part of the duodenum is inaccessible to endoscopy.

Patient preparation

Ultrasound or CT must be performed prior to PTC. These modalities provide information relating not only to the level and cause of bile duct obstruction, but also to the assessment of tumour resectability and planning of the most appropriate approach to surgical or nonsurgical biliary decompression. (Gibson *et al.*, 1986; Reiman *et al.*, 1987).

A preliminary clotting profile and platelet count should be normal or only slightly outside the normal range. Patients with a coagulopathy should receive parenteral vitamin K and blood products as appropriate.

An intravenous line is inserted and, in view of the high incidence of bacterial colonization of obstructed biliary systems (Keighley, 1977), broad-spectrum antibiotics are administered (e.g. piperacillin, 2 g, intravenously, 1 hour prior to PTC).

Table 2.1 *Comparison of PTC and ERC. (Reprinted with permission of the publishers, Gibson 1988)*

	PTC	**ERC**
Advantages	Less expertise needed. Good duct filling above an obstruction.	Visualization of stomach and duodenum. Biopsy of periampullary lesions possible. Simultaneous pancreatogram.
	Both may be followed by catheter or endoprosthesis insertion for biliary drainage.	
Contraindications	Significant coagulopathy. Marked ascites*	Unfavourable anatomy. Pseudocyst* Recent acute pancreatitis*
Success rates	98% – dilated ducts 70% – undilated ducts (Harbin et al., 1980)	Up to 90% (Cotton 1977)
Major complications	4.1% (Gibson 1988)	2–3% (Cotton 1977)
Mortality	0.13% (Gibson 1988)	0.1–0.2% (Cotton 1977)

* Relative contraindications

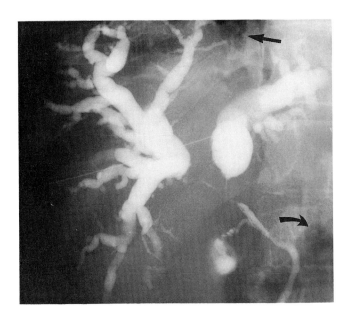

Fig. 2.1
Percutaneous transhepatic cholangiography performed from the right flank approach in a patient with hilar cholangiocarcinoma involving first-order right and left hepatic ducts. The skin puncture site is chosen under fluoroscopy and should be approximately midway between the visible lateral costophrenic angle and the lower margin of the right lobe of liver on quiet respiration. The 22 gauge Chiba needle is inserted on suspended respiration in the direction illustrated, aiming at a point approximately midway between the right cardiophrenic angle (arrow) and the gas in the first part of the duodenum or pyloric gastric antrum (curved arrow). In this case the left ducts were opacified by advancing the needle from the right. An alternative is to puncture the left lobe via the epigastrium (see Fig. 2.18b). Note that part of the left duct system usually lies anterior to the spine, so that if no opacified ducts are seen overlying the spine it usually indicates nonfilling or incomplete filling of the left hepatic ducts. Reproduced from Gibson (1988) with permission of the publishers.

Technique

Percutaneous transhepatic cholangiography is performed as a sterile procedure with the patient on a fluoroscopic table which preferably is tilting. Local anaesthesia is used with intravenous sedation and analgesia with appropriate patient monitoring.

A 15 cm long 22 gauge (0.7 mm) flexible Chiba needle is inserted, usually from the right side. The puncture site is slightly anterior to midway between the table top and the xyphisternum; inferior to the right lateral costophrenic angle on fluoroscopy on full inspiration and superior to the hepatic flexure of the colon on full expiration. The needle is inserted medially through the liver whilst screening, angulated slightly anteriorly to the coronal plane and directed craniad towards a point midway between the right cardiophrenic angle and the first part of the duodenum which can usually be identified by luminal gas (Figs 2.1 and 2.2). Specific vertebral bodies are not appropriate for guidance as their relationship to the liver is variable. Needle movement should be during suspended respiration, end-expiration or end-inspiration, and the different phases of respiration can be used to help direct the needle.

The needle tip is advanced to approximately over the right margin of the spine and then withdrawn incrementally with intermittent suction applied using a syringe connected via a short tubing. Needle entry into a bile duct is identified by aspiration of bile (which may be clear with high-grade obstruction, i.e. 'white' bile) or injection of contrast medium. If there is intrahepatic biliary dilation two or three passes using aspiration should be made before using

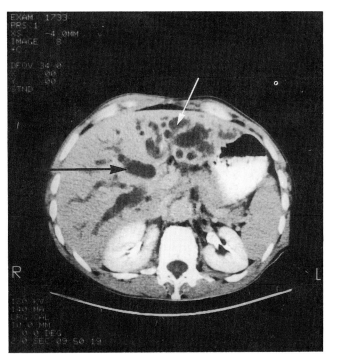

Fig. 2.2
Computed tomographic scan in a patient with marked intrahepatic duct dilation and moderate left lobe atrophy due to hilar cholangiocarcinoma. The puncture site for percutaneous transhepatic cholangiography for a right flank approach is slightly anterior to midcoronal plane and the needle is passed horizontally or slightly anteriorly (black arrow) and angled slightly craniad. In hilar bile duct obstruction, the left lobe can be punctured from the right side by angling more anteriorly as the needle is advanced to the midline, or from the epigastrium (white arrow).

contrast medium injection to identify ducts since most dilated systems will be entered with three passes or less. In the absence of biliary dilation contrast medium injection is used rather than aspiration. Injection of contrast medium outside the bile ducts should be kept to a minimum as it tends to obscure the region of interest, may be painful particularly around the hepatic hilum and within the peritoneum, and can produce pseudo-obstruction of intrahepatic bile ducts. Injection into portal or hepatic veins is frequent and is easily recognized by rapid flow of contrast medium away from the needle tip. Injection into a bile duct is identified by a characteristic slow 'oil-like' flow of contrast medium away from the needle tip. With multiple needle passes it is common to produce haemobilia. When this occurs blood-stained bile is aspirated which is more viscous than frank venous blood and drips like oil rather than water. Occasionally lymphatics are opacified and these have a characteristic fine beaded appearance (Fig. 2.3).

The number of needle passes which may be needed to opacify ducts is determined mainly by the degree of intrahepatic duct dilation. With dilated ducts the opacification rate is approximately 95% if the number of needle passes is limited to six, and approaches 100% with more passes. With undilated systems the success rate improves with the number of needle passes; there is no set limit on the number but the yield from more than 15 passes is negligible. Adequate patient analgesia is more important when ducts are nondilated and multiple needle passes are required.

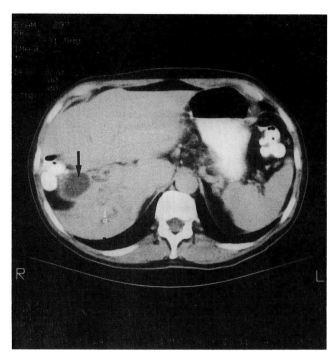

Fig. 2.4
Computed tomographic scan of a patient with marked right lobe atrophy (the right lobe lies entirely posterior to the midcoronal plane). The gall bladder (arrow) and a segment of opacified colon lie laterally. Ultrasound guidance for percutaneous transhepatic cholangiography is valuable in such a case to puncture the right lobe selectively and with safety.

For most cases fluoroscopic guidance is adequate but ultrasound guidance is useful for segmental duct obstruction or where liver anatomy is unusual, for example, following liver resection or in the presence of lobar atrophy (Fig. 2.4).

Insertion of the PTC needle via the epigastrium is preferable to a right flank approach in the following situations:

1 If only a left lobe cholangiogram is required or if a right-sided PTC has failed to produce a left lobe cholangiogram despite turning the patient on to the left side. The left lobe can be entered via a right flank approach but it is usually easier to use an epigastric approach. The left lobe lies anteriorly so if it is punctured from the right side the needle needs to be directed more anteriorly.
2 If there is right lobe atrophy or previous right hepatectomy which results in the gall bladder and/or bowel lying deep to the right lateral abdominal wall where they are at risk of puncture with a right flank approach (Fig. 2.4).

Following successful duct entry, bile samples should be obtained for bacteriological and, if malignant obstruction is suspected, for cytological examina-

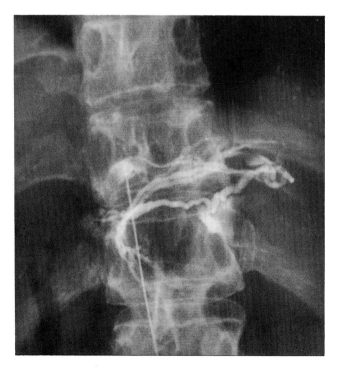

Fig. 2.3
Percutaneous transhepatic cholangiography performed via an epigastric approach opacifying numerous lymphatics which are typically fine, beaded and tortuous.

tion. Water-soluble contrast medium (200–300 mg iodine per ml) is then injected in sufficient quantities to obtain as much filling as possible of the intra-hepatic and extrahepatic systems without using un-due pressure. As much bile as possible is aspirated but if aspiration is difficult without losing needle position then full strength contrast medium is used and the bile itself will produce some dilution. In very dilated systems or if stones are suspected more dilute contrast medium is needed so that stones or ductal anatomy are not obscured.

With the needle in place the patient can be care-fully moved into a left side down oblique position which helps fill the more anterior left lobe ducts and feet down table tilt is used to fill the extraheptic bile ducts completely. Films are processed and checked prior to needle withdrawal and then further films are taken in different projections to ensure complete visualization of the biliary system. A prone film may be needed to fill the left ducts.

PITFALLS

1 *False localization of level of obstruction*
Failure to inject sufficient contrast medium or use table tilt and patient positioning can lead to false localization of obstruction above the true level. This is recognized by the presence of a 'hazy' margin at the level of apparent obstruction (Fig. 2.5).

2 *Incomplete cholangiogram*
Opacification of only the right-sided ducts is often mistaken for a complete cholangiogram. The left-sided ducts usually lie over the spine and if they fail to opacify despite patient rotation and table tilt then a separate left duct puncture is performed via either the right side or the epigastrium (Fig. 2.1).

With hilar duct obstruction not only may there be separation of the right and left hepatic ducts, but extension of stricturing into second-order ducts, particularly on the right side, can lead to non-opacification of obstructed segments. This is avoided by recognizing the normal pattern of segmental duct branching as well as the normal variants, some of which are common (Figs 2.6 and 2.7; Table 2.2).

Table 2.2 *Segmental nomenclature, after Couinaud (1957), and Healey and Schroy (1953). (Reprinted with permission of the publishers, Gibson 1988).*

I	Caudate lobe
II	Left lateral superior segment
III	Left lateral inferior segment
IV	Left medial segment or quadrate lobe
V	Right anterior inferior segment
VI	Right posterior inferior segment
VII	Right posterior superior segment
VIII	Right anterior superior segment

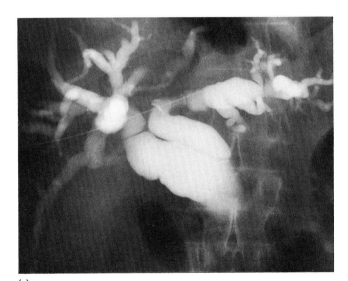

(a)

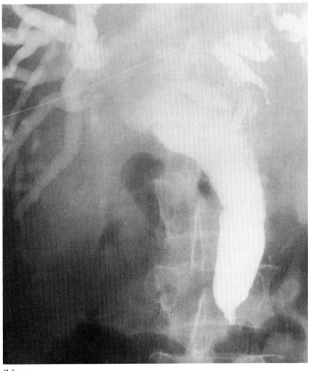

(b)

Fig. 2.5 *Percutaneous transhepatic cholangiography in a patient with low common bile duct obstruction due to ampullary carcinoma*
(a) Initially the level of obstruction appears to be in the high common duct but the hazy margin indicates that this duct is incompletely filled. (b) The true level of obstruction is demonstrated on feet-down tilt of the table.

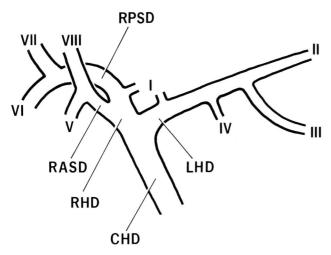

Fig. 2.6
Standard intrahepatic ductal anatomy. Segments numbered according to Couinaud's description (see Table 2.2).
RPSD = right posterior sectoral duct
RHD = right hepatic duct
CHD = common hepatic duct
RASD = right anterior sectoral duct
LHD = left hepatic duct
Reproduced from Gibson (1988) with permission of the publishers.

PERCUTANEOUS BILIARY DRAINAGE

Percutaneous biliary catheterization is used to provide access for:

1 Transhepatic biliary catheter drainage
2 Transhepatic placement of biliary endoprostheses (Chapter 3)
3 Transhepatic treatment of benign biliary strictures and/or duct stones (Chapter 4)
4 Transhepatic assistance for endoscopic procedures (Chapter 6)
5 Transhepatic placement of iridium for brachytherapy treatment of malignant strictures (Chapter 6)
6 Transjejunal treatment of benign or malignant strictures, or stone removal (Chapter 4)

Routine preoperative percutaneous biliary drainage has been abandoned in most centres as it does not lower the overall perioperative morbidity and mortality. Early retrospective studies suggested that a period of preoperative PBD reduced postoperative mortality in jaundiced patients (Nakayama *et al.*, 1978), but subsequent prospective randomized studies failed to show any advantage (McPherson *et al.*, 1984).

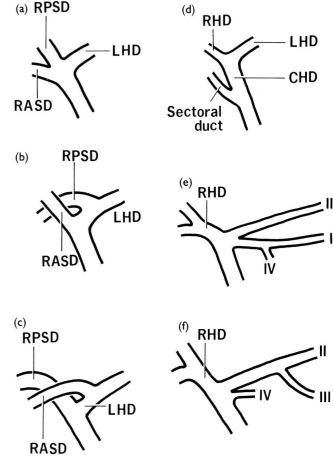

Fig. 2.7
(a–f) Variations of perihilar ductal anatomy. Segments are numbered according to Couinaud's description (see Table 2.2). Reproduced from Gibson (1988) with permission of the publishers.

This section concentrates on the technical aspects of transhepatic biliary catherization and drainage, particularly for malignant biliary obstruction.

Patient selection and preparation

Patients with malignant biliary obstruction need careful clinical and radiological evaluation to allow the most appropriate choice between surgical, radiological or endoscopic management, or a combination of approaches. Assessment of suitability for a surgical approach requires evaluation of the presence of general medical risk factors (see also Chapter 6), and detailed radiological evaluation of tumour resectability (Gibson *et al.*, 1986; Reiman *et al.*, 1987; Freeny *et al.*, 1988).

In patients with tumour which is unresectable because of either tumour extent or medical risk factors, the questions to address are:

1 *Is biliary decompression indicated?*
Most patients require decompression for troublesome symptoms of pruritus or cholangitis, and many patients want decompression to treat the jaundice itself. There are, however, some patients who have minimal symptoms apart from jaundice, or who have a very limited prognosis in whom any sort of intervention is meddlesome. Patients with multiple obstructed and isolated biliary segments, for example, due to liver metastases or extensive hilar cholangiocarcinoma, in general should not undergo PBD.

2 *Is surgical or nonsurgical decompression more appropriate?*
This decision has often been made prior to referral for radiological intervention. Surgical decompression is more appropriate if there is gastric outlet obstruction. The use of surgery to make a diagnosis is usually unnecessary and appears to increase the morbidity of subsequent nonoperative biliary drainage (Gibson *et al.*, 1988).

3 *Is percutaneous or endoscopic intervention more appropriate?*
Providing there is endoscopic access to the second part of the duodenum this approach is generally the first choice for stenting of mid and low common bile duct lesions. For hilar ductal obstruction the choice is less clear and is influenced by local expertise. With hilar obstructions extending into first- or second-order right or left hepatic ducts it is important to select the most appropriate lobe or segment for stent placement. If this cannot be achieved with an endoscopic approach the endoscopist should withdraw rather than leave a stent in an inappropriate lobe or segment as this often leads to cholangitis in a segment which otherwise would not need drainage. Selection of a specific lobe or segment is more reliably achieved with a percutaneous approach, either alone or as part of a combined percutaneous–endoscopic approach (Chapter 6).

Cytological or histological confirmation of suspected malignant obstruction should be obtained. This is particularly relevant in differentiating between carcinoma of the pancreatic head and chronic pancreatitis, and in cases of possible lymphoma. Percutaneous needle aspiration or biopsy can be performed using cholangiographic guidance under fluoroscopy (Fig. 2.8), or using ultrasound or CT if there is a demonstrable mass (Hall-Craggs and Lees, 1985). Obtaining a positive diagnosis of malignancy may be difficult in some situations, particularly with cholangiocarcinoma, and brushings or biopsy via a percutaneous catheter

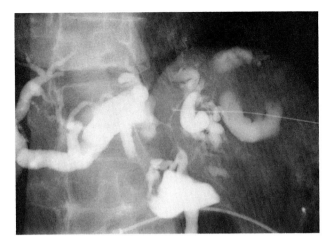

Fig. 2.8
Fine needle aspiration using cholangiographic guidance in a patient with hilar cholangiocarcinoma. Sampling in this case is from the left hepatic duct stricture.

track can be used.

In addition to addressing the above questions, patient preparation should include the following:

1 Clotting profile and platelet count should be normal or only slightly outside the normal range, and any defect corrected with appropriate blood products. Parenteral vitamin K should be administered routinely if possible for 2 to 3 days prior to PBD in view of the likelihood of disturbance of vitamin K-dependent clotting factors in patients with bile duct obstruction.

2 Attention should be paid to fluid and electrolyte balance. Many patients are significantly dehydrated which can contribute to renal failure, and intravenous rehydration is frequently warranted.

3 Adequate imaging should be undertaken to determine the most appropriate approach to biliary decompression. This should include ultrasound and usually CT, and for high obstructions direct cholangiography (PTC or ERC). Specific factors to assess are:

(a) The proximal extent of bile duct stricturing. This may be evident on ultrasound or CT, although for hilar obstruction direct cholangiography is essential for an accurate assessment (Gibson *et al.*, 1986). The proximal extent of stricturing is much more important than the lower extent, and it is essential that this be established accurately for strictures close to, or involving, the confluence of the right and left hepatic ducts. This information is vital in determining resectability of hilar tumours and, in those which are unresectable, in selecting the most appropriate means of palliative biliary decompression (see below).

(b) The presence of intrahepatic tumour. Either

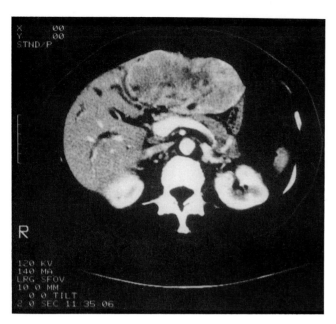

Fig. 2.9
The presence of extensive tumour in one hepatic lobe (in this case the left) dictates that biliary drainage should be via the contralateral lobe.

ultrasound or CT will determine if there is extensive liver tumour in one lobe which may determine whether a right or left lobe approach is more suitable (Fig. 2.9).
(c) The presence of lobar liver atrophy. Computed tomography is superior to ultrasound in detecting lobar atrophy which may be present with hilar obstruction, particularly due to

cholangiocarcinoma (Carr *et al.*, 1985) and results from prolonged lobar bile duct obstruction and/or portal vein branch involvement by tumour. Lobar atrophy is recognized by crowding of dilated bile ducts in a lobe which is usually appreciably reduced in size, often with contralateral lobar hypertrophy (Fig. 2.10). Lobar atrophy more commonly affects the left lobe.
(d) The presence of intrahepatic abscesses. Computed tomography is the modality of choice for detection of abscesses which should be suspected if there is clinical sepsis.
(e) The presence of ascites. Ascites is a relative contraindication to PBD and should be noted on ultrasound or CT.

Broad-spectrum antibiotic cover is used routinely in view of the high incidence of bacterial colonization of the obstructed biliary system with gut flora. A suitable dose is given prior to the procedure to allow adequate tissue levels to be achieved, for example, piperacillin, 2 g, given intravenously 1 hour prior to the procedure. In patients who are very unwell from sepsis, antibiotic cover for anaerobic organisms should be added (e.g., metronidazole). Antibiotic therapy should be modified according to culture and sensitivities if positive bile or blood cultures have been obtained.

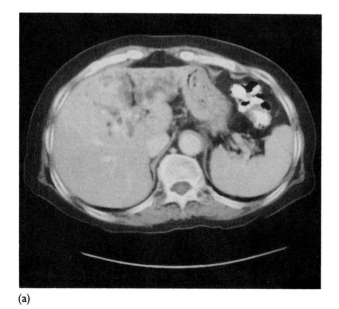

(a)

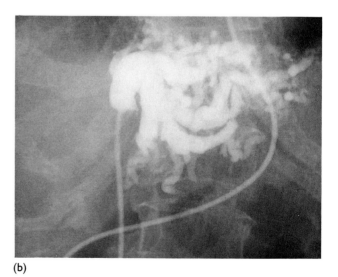

(b)

Fig. 2.10 *Hilar cholangiocarcinoma associated with left lobe atrophy*
(a) The computed tomogram and (b) the nasobiliary cholangiogram show that the left lobe ducts are more dilated than those on the right, and are very crowded, with a corresponding reduction in hepatic parenchymal volume of the left lobe.

Contraindications

1 Uncorrectable coagulopathy
2 Marked ascites
3 Unsafe access route. This is unusual but occasionally applies, for example, if colon interposes between right lobe of liver and right flank and a left lobe approach via the epigastrium is either not appropriate or not possible.
4 Numerous obstructed and noncommunicating intrahepatic biliary segments.

Technique

The procedure is performed with sterile technique under fluoroscopic guidance, and if biplane or C-arm fluoroscopy is available it can assist selective duct puncture and reduce screening time. Single-plane screening, however, is usually quite adequate with the patient being turned obliquely as needed. Ultrasound guidance is used routinely by some, but it is most useful in patients with segmental duct obstruction or where liver anatomy is unusual, for example, following liver resection or in the presence of lobar atrophy (Fig. 2.4).

Most cases can be performed with intravenous analgesia and sedation provided that adequate doses are used which in turn demands close monitoring of respiratory and circulatory status, and the use of a pulse oximeter is desirable. The assistance of an anaesthetist is helpful and general anaesthesia is necessary in some patients.

CHOICE OF APPROACH

Percutaneous biliary drainage can be performed via the right flank for entry into a right lobe duct, or via the epigastrium for a left lobe puncture. For obstruction to the common duct a right-sided approach is prefered as it provides a straighter approach to the common duct (the left hepatic duct tends to have a more acute angle with the common duct), and it is easier for the operator to keep their hands out of the primary X-ray beam.

For hilar duct obstruction where there is partial or complete separation of the right and left hepatic ducts it is important to use CT as well as the cholangiogram to assess carefully the proximal extent of stricturing and the presence of lobar atrophy. The choice of approach may be decided by the presence of atrophy, extensive tumour, or multiple isolated segments in one lobe; in these situations the contralateral lobe should be drained (Fig. 2.11). If clear bile ('white' bile) is aspirated from one lobe and

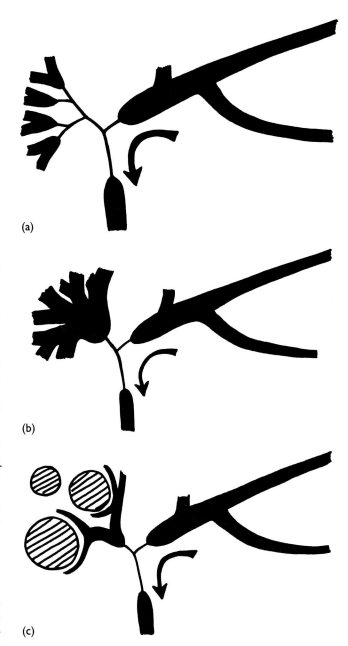

(a)

(b)

(c)

Fig. 2.11
Factors affecting the choice of right versus left lobe approach to biliary drainage. The presence in one lobe of (a) stricturing extending into third- or fourth-order hepatic ducts producing multiple isolated segments, (b) significant atrophy, or (c) extensive tumour, dictates that drainage should be of the contralateral lobe.

'yellow' bile from the opposite lobe, the latter lobe should be drained preferentially, other factors being equal. 'White' bile indicates longstanding obstruction to that lobe which therefore is more likely to have undergone significant hepatocellular injury or lobar atrophy.

In addition, in patients with clinical cholangitis attempts should be made to aspirate bile from each isolated lobe or segment using the Chiba needle. If

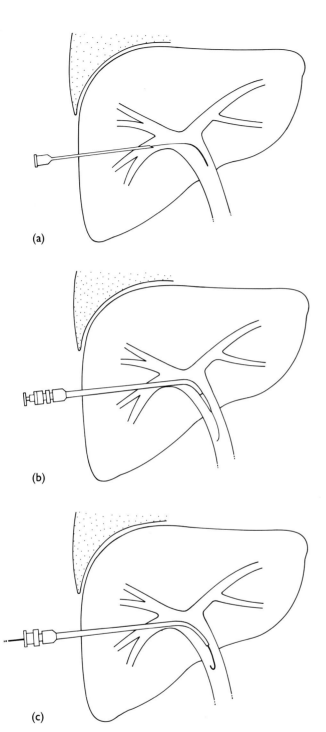

Fig. 2.12 *Percutaneous biliary drainage: duct puncture and guidewire introduction*
(a) The selected duct is punctured with a 22 gauge needle and, after confirming position, a 0.018 inch guidewire is inserted. (b) A Neff introduction system is passed over this wire and the outer sheath left in place, through which is inserted a 0.038 inch J-wire. (c) A biliary manipulation catheter can then be passed over the J-wire.

one lobe or segment yields purulent bile then this lobe or segment should be drained preferentially.

The aim of biliary drainage is to drain as much tumour-free and nonatrophic liver as possible. Fewest errors will be encountered if a complete cholangiogram is obtained prior to selection of the most appropriate duct for puncture. The author's preference is to use a right lobe approach unless contraindicated on any of the above grounds (Fig. 2.11).

Right lobe approach (Fig 2.12 and 2.13)
The skin puncture site is similar or slightly posterior to that used for PTC (see above) and is modified appropriately to provide the most direct line of approach to the duct selected for puncture. The aim is to choose a duct which allows a line of approach to the common bile duct without acute angles. It is important to minimize angles in all planes. A branch of the right posterior duct is usually suitable unless this drains into the left hepatic duct (Fig. 2.7b); furthermore, the right posterior duct usually drains a larger volume of liver than the right anterior duct, which is of importance with hilar duct obstruction.

In selecting the intrahepatic duct for puncture it is important to study the cholangiogram for variations in the intrahepatic pattern of bile duct branching (Fig. 2.7). The commonest variations of practical importance are those in which the right posterior or right anterior sectoral duct drains into the left hepatic duct; this occurs in 22 and 6% of patients, respectively (Figs 2.7b and c) (Healey and Schroy, 1953). If either of these ducts is catheterized it creates an 'S-bend' approach to the common duct making manipulation through strictures difficult. This can be avoided by obtaining a complete cholangiogram before selecting the duct to be catheterized.

After opacifying the intrahepatic ducts the selected duct is punctured a few centimetres away from the hilar region to reduce the risk of injury to the larger portal veins and hepatic arteries in the perihilar region (Figs 2.12a and 2.13a). A 22 gauge Chiba needle is used for the initial puncture. Puncture of the duct under screening control is recognized by distortion of the duct wall and usually a perceptible 'give' as the needle enters the duct. The antero-posterior relationship between the needle and the duct can be assessed by partial rotation of the patient if C-arm or biplane fluoroscopy is not available. The needle can be withdrawn either partially or totally and redirected as appropriate. Intraductal location is confirmed by aspiration of bile (which may be bloodstained but will still drip like oil) and then injection of contrast medium. It is preferable to use aspiration prior to injection as this avoids obscuring

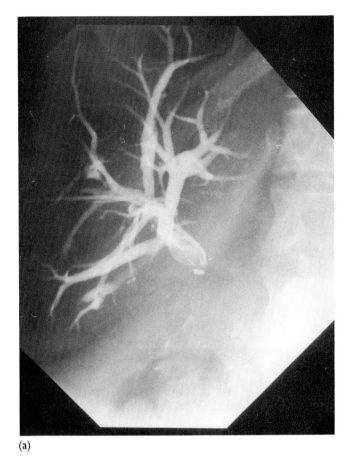

(a)

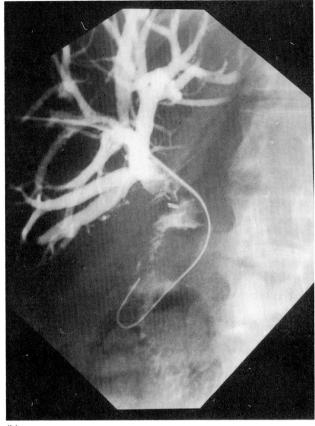

(b)

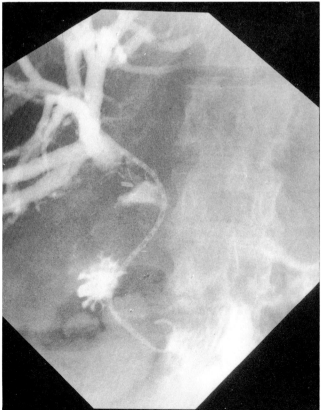

(c)

Fig. 2.13 *Internal/external percutaneous biliary drainage in a patient with pancreatic carcinoma*
(a) A biliary manipulation catheter has been introduced to the level of obstruction. (b) A straight wire is passed through the stricture, and over this a drainage catheter is introduced. (c) This is usually preceded by dilation of the tract. In this case a Ring catheter has been positioned so that side holes lie above the stricture and the distal end lies in the duodenum, establishing internal and external drainage.

the region of interest with intraparenchymal injections of contrast medium. The needle will tend to occlude with clot and should then be removed and flushed thoroughly, and the needle stylet cleaned, prior to repuncture.

Once in the chosen duct an introducing system such as the Neff (Fig. 2.12b) or Cope (Cook) is used to introduce a 0.038 inch J guidewire (Fig. 2.12c). The Cope system delivers the J-wire through a side hole and the Neff system through the end hole. Both systems work well in large ducts but the Neff system is prefered for smaller ducts, and in general is easier to use. If the 0.018 inch guidewire will not advance freely into the duct it can be withdrawn into the needle which is then advanced or withdrawn slightly under screening and the wire gently advanced again. If it still will not advance freely but its position appears good on screening it may be because a small side branch of the duct has been punctured. Further adjustment of needle position in the anteroposterior plane is then necessary.

When the outer sheath of the introducing system appears to be in the bile duct contrast medium is injected to confirm its position. Occasionally the introducing system enters the portal vein accompanying the bile duct and when this occurs the system should be withdrawn and duct puncture reattempted at a different site.

When the J-wire is in the selected duct it is advanced towards the level of obstruction and the introducing catheter is replaced with a manipulation catheter which should be a short torque-control catheter with a very short distal bend (Fig. 2.14). The combination of the manipulation catheter and J-wire can then sometimes be advanced through the stricture, although it is usually necessary to change to a straight wire. The Amplatz extrastiff wire (Cook), which has a relatively soft distal end, may be used but if the stricture is not easily entered with this wire it should be replaced with a softer tipped wire such as the Newton (Cook). The catheter is positioned just proximal to the stricture and the wire used to probe the stricture gently to find the lumen, the direction of the wire being controlled by the catheter (Fig. 2.15). Even if no contrast medium passes through the stricture it is frequently possible to pass the wire with gentle probing. It is important not to use any force on the wire or catheter until the stricture is entered as this will tend to produce a blind false passage and it will be difficult to locate the true lumen. Occasionally as last resort it is necessary to create a false lumen but if the stricture cannot readily be negotiated at the first session it is preferable to

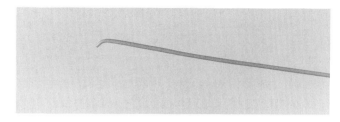

Fig. 2.14
Biliary manipulation catheter which has good torque control and a very short 45° angled distal tip (Cook, BPS 6.5-38-40-P-NS-GBM-1).

leave an external drainage catheter in place above the stricture and attempt to cross the stricture a few days later (Fig. 2.16). Frequently a complete obstruction will convert to an incomplete obstruction allowing the stricture lumen to be opacified. A hydrogel-coated wire (Radifocus wire, Terumo, Japan) can be helpful for negotiating tight tortuous strictures; this type of wire should then be replaced with a conventional wire for subsequent dilation and catheter placement. In patients with suppurative cholangitis manipulations should be kept to a minimum and if the stricture cannot be passed easily it is better to establish external catheter drainage and attempt to cross the stricture after a few days of drainage and antibiotic therapy.

When the wire is successfully passed through the stricture the manipulation catheter is withdrawn and replaced with a drainage catheter placed across the length of the stricture (Figs 2.13c and 2.17). thereby establishing internal/external drainage. It is usually necessary to dilate the liver tract and the stricture using serial or coaxial dilators depending on the size of the catheter and the tightness of the stricture.

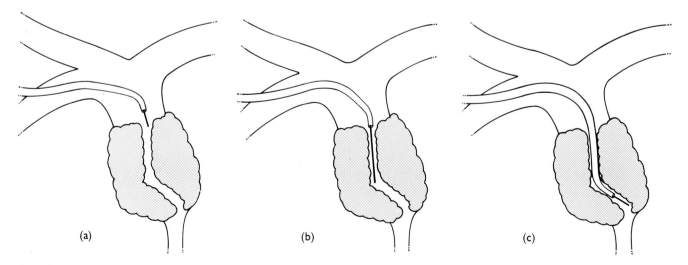

Fig. 2.15
(a,b) The combination of a biliary manipulation catheter with good torque control, and a straight wire is used to find the lumen of the stricture, and then (c) negotiate tortuous strictures.

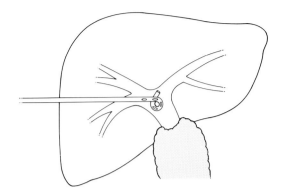

Fig. 2.16
An external drainage catheter is left in place if the obstruction cannot be negotiated. The catheter should have some intrinsic stability such as a pigtail end, with several distal side holes; contrast medium injection is used to ensure that all side holes lie within the duct.

Left lobe approach

The left lobe approach for PBD is used in preference to a right lobe approach in the situations illustrated in Fig. 2.11. Some interventional radiologists use a left lobe approach routinely, but this is not the author's preference. Applying the same principles as illustrated in Fig. 2.11 a left lobe approach should not be used if that lobe is atrophic, contains a large tumour load, or contains multiple isolated biliary segments. Drainage of both lobes is sometimes indicated for hilar duct obstruction, but it is not necessary to drain both right and left lobes simply because there is hilar obstruction (see below).

The approach for left lobe drainage is via the epigastrium as shown in Fig. 2.18. The principles of duct puncture and catheterization are the same as for a right lobe approach. Very occasionally the left lobe can be approached via the left flank if there is hilar duct obstruction and very marked left lobe hypertrophy associated with right lobe atrophy.

MULTIPLE DRAINS OR A SINGLE DRAIN IN HILAR BILIARY OBSTRUCTION?

In patients with hilar bile duct obstruction there is frequently separation of the right and left lobe bile ducts and often isolation of segmental ducts. The aim in these cases is to drain the largest segment of tumour-free and nonatrophic liver. In general it is not appropriate to perform PBD unless drainage of at least half of one lobe is achievable with one catheter. It is usually possible to produce adequate palliation with only one catheter (or endroprosthesis), and separate drainage of more than one segment or lobe only becomes necessary under the following circumstances:

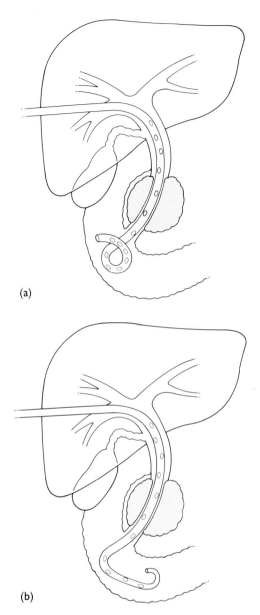

(a)

(b)

Fig. 2.17 *Internal/external biliary drainage with (a) Cope and (b) Ring catheters (Cook)*
Proximal side holes lie above the stricture and the distal end is placed in the duodenum for internal drainage and stability.

1 If both lobes contain frankly infected bile which is causing clinical sepsis. This may be evident at the time of initial PBD as revealed by fine needle aspiration of purulent bile from both lobes. Alternatively it may become manifest some time after the initial PBD with the development of cholangitis which does not settle with appropriate antibiotic therapy. Aspiration of the undrained lobe or segments may then reveal purulent bile and insertion of a second PBD catheter can then be considered.

2 If the initial catheter fails to produce any relief of jaundice and there is at least half a lobe of non-

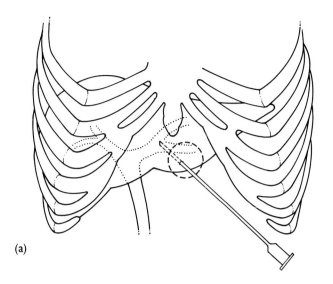

(a)

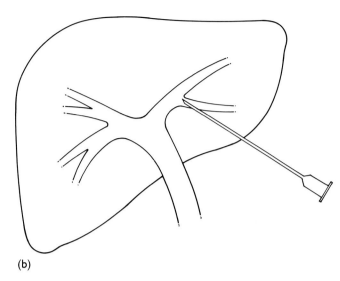

(b)

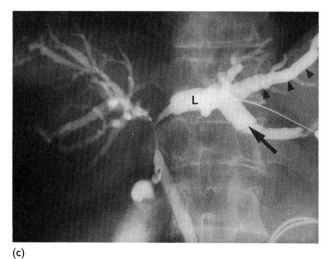

(c)

Fig. 2.18
Percutaneous biliary drainage (PBD) of the left lobe via an epigastric approach. The position and anatomy of the left lobe can be determined from prior computed tomography (CT) or ultrasound, and cholangiography at the time of PBD. (a) The skin is punctured over the inferior margin of the visible left lobe, in or to the left of the midline, to gain the most direct approach to the selected left lobe duct, which usually is the segment 3 duct (b) The segment 3 duct passes anteriorly as well as inferiorly so is preferred for an epigastric approach. (c) Percutaneous transhepatic cholangiography in a patient with hilar cholangiocarcinoma. The arrow indicates the approach to the segment 3 duct for left lobe PBD. The segment 2 duct (arrowheads) passes posteriorly and superiorly which creates unfavourable angles for an anterior approach. If the segment 3 duct cannot be punctured the next best option is usually the main left hepatic duct (L).

atrophic tumour-free liver which remains undrained.

The routine use of more than one PBD catheter in order to drain multiple lobes or segments increases patient discomfort and almost certainly increases the early complication rate with no proven significant benefit in terms of duration or quality of survival.

CHOICE OF DRAINAGE CATHETER AND POSITIONING

The size and type of drainage catheter chosen, and its positioning, depend on several factors.

Size
If biliary drainage is short term an 8 Fr catheter will usually suffice, but for longer term drainage 10 Fr is preferable. For very short term (1 or 2 days), for example, in performing 'rendezvous' procedures (see Chapter 6), a 6 to 7 Fr catheter is adequate. To some extent the size is determined by the ease of insertion.

In patients managed by internal/external drainage catheters, after 1 to 2 weeks of drainage it is worth attempting conversion to internal drainage by capping the open end of the catheter. Sometimes this is followed by leakage of bile externally around the catheter in which case increasing the catheter size to 12–14 Fr may control the external leakage by improving internal drainage.

Shape
A variety of shapes are available most of which incorporate some type of intrinsic stability (Fig. 2.19). For *internal/external drainage*, catheters such as the Cope or Ring designs (Cook) are suitable. (The Ring catheter is more easily inserted whereas the Cope catheter is more comfortable for long-term drainage.) For *external drainage* catheter choice is determined by the level of obstruction and the

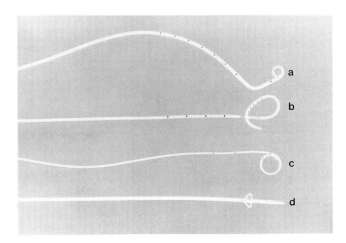

Fig. 2.19 *Selected biliary drainage catheters*
(a) Ring biliary (Cook). (b) Cope biliary (Cook). (c) Cope nephrostomy (Cook). (d) Castaneda nephrostomy (Meditech). (a) and (b) are ideal for internal/external drainage. (c) and (d) are useful for placing above the obstruction for external drainage.

degree of duct dilation. In mid to low common duct obstructions a pigtail catheter such as the Cope nephrostomy or Kumpe catheter (Cook) is suitable. For high common duct and hilar obstructions there is usually insufficient space for a pigtail, although for short-term drainage a small diameter pigtail (e.g. 6.5 Fr Cope nephrostomy, Cook) can usually be accommodated. For longer term drainage a catheter with a Mallecot-type anchor (Fig. 2.19) is suitable. If a straight catheter is left in place for very short-term external or internal drainage, a good length of catheter must be inserted. For external drainage this means manipulating the catheter well into the other lobe or another segment of the liver.

Side holes and positioning
Several side holes should lie above the level of the obstruction for both internal and external drainage, and for internal drainage a few side holes should lie below the lower end of the stricture. Prior to inserting the drainage catheter the length of the biliary stricture is measured using the guidewire inserted through the stricture. Extra side holes may need to be made using a hole punch. When the catheter is placed, contrast medium is injected under screening to check that:

1 There are side holes above the stricture. If the ducts above the stricture do not opacify, the catheter is gradually withdrawn under screening whilst contrast medium is injected until there is free duct opacification. Extra proximal side holes are made if necessary.
2 There is no filling of vascular structures. If vascular opacification occurs the catheter should be

advanced gradually until no opacification occurs. If this cannot be achieved a separate duct puncture and drainage should be undertaken. The initial catheter may be left in place until this is completed.
3 There is no leakage of contrast medium into the peritoneal cavity. If leakage occurs the catheter should be advanced until leakage stops. If leakage continues this generally means that the side holes extend too far proximally and either a different catheter with fewer side holes should be used or a more peripheral duct puncture performed so that there is more catheter within the duct.

When the obstruction lies at the lower end of the common bile duct the lower end of an internal/external drainage catheter is positioned, of necessity, in the duodenum. With higher strictures the lower end is also best placed in the duodenum rather than in the lower common bile duct as this helps secure the catheter.

Catheter material
The catheter should be moderately flexible to reduce discomfort and to allow bending outside of the patient, and should not kink permanently. All of the currently available materials are prone to occlude with biliary encrustations.

PROBLEMS ON CATHETER INSERTION

Guidewire kinking
This is particularly prone to occur if a stiffening cannula is used such as with the Cope drainage catheter. The risk of kinking is reduced by keeping the stiffening cannula a short distance back from the first bend in the course of the guidewire as it approaches the hepatic hilus. Once a guidewire kinks it should be replaced.

Catheter looping outside the liver
This can occur during advancement of the catheter over the guidewire when the catheter meets resistance, usually at the biliary stricture. The tendency to loop is reduced by:

1 Initially selecting an approach which avoids angles or tight curves
2 Preliminary stricture dilation
3 Using a stiff guidewire (e.g. Amplatz extrastiff)
4 Keeping the guidewire straight and advancing the catheter off it rather than advancing both catheter and wire together
5 Using a stiffening cannula such as with the Cope set

Catheter will not advance through the stricture
This can usually be corrected by one or more of the following:

1 Dilating the stricture further
2 Using a stiff guidewire (e.g. Amplatz extrastiff) or stiffening cannula
3 Changing to a smaller diameter catheter
4 Changing to a stiffer catheter, such as a Ring catheter, or as a temporary catheter a van Andel (Cook). The latter is straight and is therefore not suitable for long-term drainage because of the risk of dislodgement. It may, however, allow a track to develop so that a stable catheter can be introduced after 2 to 3 days.

FIXING THE CATHETER

A variety of methods exist for fixing the catheter to the skin. The fixation should be secure and comfortable for the patient. A simple means of achieving both requirements is to use the method shown in Fig. 2.20. The stomal wafer must be changed regularly together with attention to local skin care. Skin sutures should be avoided if possible as they produce significant discomfort.

Postprocedural care

IMMEDIATE

Initially the drainage catheter is left to drain externally whether it lies above or through the obstruction. The patient is observed carefully for at least 24 hours and adequate analgesia maintained. The catheter is flushed in the ward and flushing instructions should be simple and clear, for example, 'inject 10 ml sterile normal saline through the catheter 6 hourly for 48 hours'. The volume of bile drained is charted daily, and fluid and electrolyte balance maintained. External biliary drainage has the potential to produce significant dehydration and electrolyte loss. Antibiotics are continued after drainage catheter insertion if the patient has any signs of sepsis, and modified as appropriate according to bile culture and sensitivities.

If an endoprosthesis is being inserted as a two-stage procedure this is performed usually after 3 to 5 days of catheter drainage. Patients who are being discharged with the catheter in place are instructed in the care of the catheter and drainage bag, and domiciliary nursing care arranged as necessary.

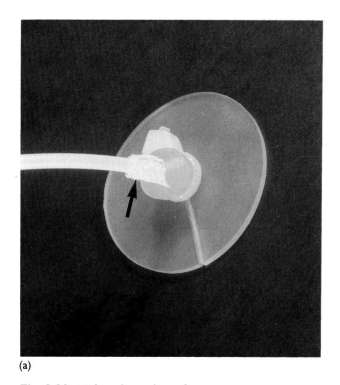

(a)

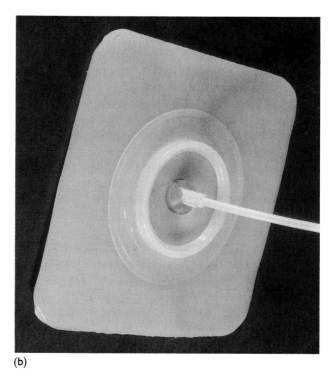

(b)

Fig. 2.20 *Molnar disc catheter fixation*
(a) A 1 cm² piece of sticking plaster (arrow) is first applied to the catheter. This provides good friction between the disc and the catheter.
(b) An adhesive stomal wafer secures the disc to the skin and a stomal drainage bag can then be fixed to the wafer.

LONG TERM

Patients are generally reviewed after 2 to 3 weeks at which time a repeat attempt to cross a biliary obstruction may be successful if this has not already been achieved. Patients with internal/external catheters can have these converted to internal drainage by capping of the catheter, and told to uncap it if bile subsequently leaks around the catheter. If patients cannot cope with the catheter, conversion to an endoprosthesis is indicated.

Catheter problems

BILE LEAKAGE AROUND THE CATHETER

A cholangiogram is performed via the catheter and the following action taken.

1 Change catheter if blocked
2 Check that there are no side holes lying peripherally in the parenchymal tract
3 If catheter position appears satisfactory change to a larger size to promote internal drainage
4 Check that there is no obstruction to the internal drainage of bile, for example, due to duodenal compression by tumour

These patients may already have had a gastroenterostomy to relieve gastric outlet obstruction. Changing the catheter for a longer catheter with the distal end lying in the third part of the duodenum or beyond may be helpful.

CATHETER OCCLUSION

This is inevitable and it is preferable to change catheters routinely at 1 month or, at most, 2 month intervals.

CATHETER DISLODGEMENT

A replacement catheter should be inserted down the tract as soon as possible to avoid the need for repeat transhepatic puncture and drainage. Choice of a catheter with intrinsic stability and good skin fixation (see above) usually prevents dislodgement.

Complications

SUBPHRENIC ABSCESS

It is common for some bile to accumulate in the extrahepatic space after the initial drainage. If the patient has any signs of sepsis the collection should be aspirated and then drained by catheter if frankly infected. Collections do not need drainage if they are not infected.

INTRAHEPATIC ABSCESS

Abscesses may develop in the liver secondary to cholangitis. Small abscesses usually respond to antibiotic therapy particularly if they communicate with decompressed bile ducts. Larger abscesses not draining via bile ducts will not resolve without percutaneous catheter drainage.

CHOLANGITIS

Cholangitis is generally controlled by antibiotics provided that the infected bile is draining. Persistent cholangitis generally means the catheter is not draining adequately and needs attention (see above) or that there is an isolated lobe or segment which is infected. Fine needle aspiration of undrained segments is attempted and if this yields purulent bile additional percutaneous drainage should be considered, taking into account the overall clinical status of the patient.

BLEEDING

Bleeding at the initial drainage is usually minor and settles spontaneously. More significant bleeding rarely requires embolization, which is transarterial for hepatic artery branch bleeding, and transhepatic for portal vein bleeding.

OTHER COMPLICATIONS

A number of other complications can occur but are very uncommon. The pleural space is commonly crossed in procedures performed from the right flank but it is surprisingly uncommon to develop a pneumothorax or pleural collection. Gall bladder puncture is more likely to occur if the gall bladder lies to the right (e.g. in right lobe atrophy, Fig. 2.4) but it is readily recognizable and, provided that bile duct decompression is achieved, does not usually warrant separate catheter drainage. If the hepatic flexure of the colon is punctured the catheter should be withdrawn, a separate puncture made for biliary drainage, and the patient observed; if peritonitis or a collection develops these are treated as appropriate. Duodenal perforation by the distal end of the internal drainage catheter has been described; this is rare and is treated on clinical merit.

Results

External catheter drainage can be achieved in virtually 100% of patients and internal catheter placement in 80–90%, although approximately 20% of these patients fail to lose their jaundice (Dooley *et al.*, 1984; Mueller *et al.*, 1985).

The 30-day mortality figures for PBD range from 10 to 30%. Median survival for PBD for malignant disease is 3–9 months with most patients surviving 6 months or less; there are, however, relatively long-term survivors in most series (Leung *et al.*, 1983; Dooley *et al.*, 1984; Mueller *et al.*, 1985; Passariello *et al.*, 1985; Lammer and Neumayer, 1986; Gibson *et al.*, 1988).

Iridium therapy via percutaneous catheters

Internal irradiation of malignant biliary strictures using iridium-192 wires can produce good palliation in some patients, particularly those with small tumour mass, usually cholangiocarcinoma (Fletcher *et al.*, 1981). The wire can be delivered via a naso-biliary catheter (see Chapter 6) or via other routes, such as:

1 A transhepatic catheter
2 Via a superficially fixed Roux loop (Fig. 2.21). This is possible if a biliary-enteric anastomosis has been performed with superficial fixation of the Roux loop (see Chapter 4)
3 Via an operatively placed tube such as a T-tube

A catheter of sufficient diameter to carry the iridium wire is placed across the malignant sticture and a 'dummy' iridium wire inserted through it and carefully positioned at the optimal site for irradiation. The delivery catheter is then securely fixed so that the iridium wire can subsequently be placed in precisely the same position for internal irradiation.

Percutaneous catheter drainage or endoprosthesis for malignant biliary obstruction?

The author's preference is to use endoprostheses rather than percutaneous catheters for long-term decompression of malignant biliary obstruction. The *disadvantages* of a percutaneous drainage catheter compared with an internal endoprosthesis include local skin discomfort, the need for daily care of the catheter and drainage bag, and the constant reminder

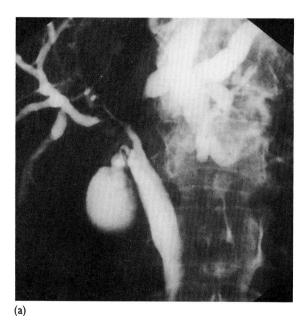

(a)

Fig. 2.21
(a) Patient with unresectable hilar cholangiocarcinoma. (b) A Roux loop has been anastomosed to the common duct below the stricture and its efferent limb fixed to the anterior abdominal wall. This loop was subsequently used to place iridium wires (arrowheads) percutaneously across the right and left strictured ducts.

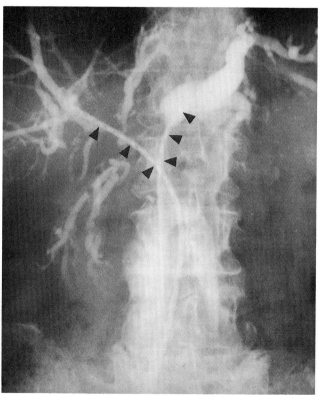

(b)

to the patient of their illness. The *advantage* of a percutaneous drainage catheter is the ready access for catheter change to maintain patency and good biliary drainage. Whilst occluded endoprostheses can be replaced percutaneously (Adam, 1987) this does involve further patient discomfort and morbidity, and is technically more demanding. Endoprostheses inserted percutaneously can usually be replaced endoscopically if the lower end of the endoprosthesis is placed in the duodenum. The reasons for selecting a percutaneous approach over an endoscopic one in the first instance may mean that endoscopic replacement is not feasible. Endoscopic replacement of an existing endoprosthesis, however, is technically easier than endoscopic insertion of the initial stent.

The choice between initial placement of a percutaneous drainage catheter and an internal endoprosthesis (see Chapter 3) is influenced by several factors, namely:

1 Patient tolerance of and ability to manage a percutaneous catheter. For most patients this will present problems and an endoprosthesis is preferable. Some patients may elect to undergo a trial of percutaneous catheter drainage which can be converted to an endoprosthesis if the catheter proves intolerable.
2 Patient prognosis. If this is very limited but palliative decompression is considered appropriate then an endoprosthesis is clearly preferable to a percutaneous catheter.

REFERENCES

ADAM A (1987). Use of the modified Cope introduction set for transhepatic removal of obstructed Carey–Coons biliary endoprostheses. *Clinical Radiology* **38**: 171–4.

CARR DH, HADJIS NS, BANKS L, Hemingway AP, BLUMGART LH (1985). Computed tomography of hilar cholangiocarcinoma: a new sign. *American Journal of Roentgenology* **145**: 53–6.

COTTON PB (1977). Progress report: ERCP. *Gut* **18**: 316–41.

COUINAUD C (1957). Le Foie. *Etudes Anatomiques et Chirurgicales.* Masson, Paris.

DOOLEY JS, DICK R, GEORGE P, KIRK RM, HOBBS KEF, SHERLOCK S (1984). Percutaneous transhepatic endoprosthesis for bile duct obstruction. Complications and results. *Gastroenterology* **86**: 905–9.

FLETCHER MS, BRINKLEY D, DAWSON JL, NUNNERLEY H, WHEELER PG, WILLIAMS K (1981). Treatment of high bile duct carcinoma by internal radiotherapy with iridium-192 wire. *Lancet* **ii**: 172–4.

FREENY PC, MARKS WM, RYAN JA, TRAVERSO LW (1988). Pancreatic ductal adenocarcinoma – diagnosis and staging with dynamic CT. *Radiology* **166**: 125–33.

GIBSON RN (1988). Percutaneous transhepatic cholangiography. In: *Surgery of the Liver and Biliary Tract*, pp. 241–55. Edited by Blumgart LH. Churchill Livingstone, Edinburgh.

GIBSON RN, YEUNG E, HADJIS N, ADAM A, BENJAMIN IS, ALLISON DJ, BLUMGART LH (1988). Percutaneous transhepatic endoprostheses for hilar cholangiocarcinoma. *American Journal of Surgery* **156**: 363–7.

GIBSON RN, YEUNG E, THOMPSON JN, CARR DH, HEMINGWAY AP, BRADPIECE HA, BENJAMIN IS, BLUMGART LH, ALLISON DJ (1986). Radiological evaluation of bile duct obstruction: level, cause and tumour resectability. *Radiology* **160**: 43–7.

HALL-CRAGGS MA, LEES WR (1985). Fine-needle aspiration biopsy: pancreatic and biliary tumours. *American Journal of Roentgenology* **147**: 399–403.

HARBIN WP, MUELLER PR, FERRUCCI JT (1980). Transhepatic Cholangiography. Complications and use patterns of the fine-needle technique. A multi-institution survey. *Radiology* **135**: 15–22.

HEALEY JE, SCHROY PC (1953). Anatomy of the biliary ducts within the human liver. *American Medical Association Archives of Surgery* **66**: 599–616.

KEIGHLEY MRB (1977). Micro-organisms in the bile. A preventable cause of sepsis after biliary surgery. *Annals of the Royal College of Surgeons of England* **59**: 328–34.

LAMMER J, NEUMAYER K (1986). Biliary drainage endoprosthesis: experience with 201 patients. *Radiology* **159**: 625–9.

LEUNG JWC, EMERY R, COTTON PB, RUSSELL RCG, VALLON AG, MASON RR (1983). Management of malignant obstructive jaundice at The Middlesex Hospital. *British Journal of Surgery* **69**: 584–6.

McPHERSON GAD, BENJAMIN IS, HODGSON HJF, BOWLEY NB, ALLISON DJ, BLUMGART LH (1984). Pre-operative percutaneous transhepatic biliary drainage: results of a controlled trial. *British Journal of Surgery* **71**: 371–5.

MUELLER PR, FERRUCCI JT, TEPLICK SK, VAN SONNENBERG E, HASKIN P, BUTCH RJ, PAPANICOLAOU N (1985). Biliary stent endoprosthesis: analysis of complications in 113 patients. *Radiology* **156**: 637–9.

NAKAYAMA T, IKEDA A, OKUDA K (1978). Percutaneous transhepatic drainage of the biliary tract. *Gastroenterology* **74**: 544–59.

PASSARIELLO R, PARONE P, ROSSI P, SIMONETTI G, MODINI C, LASAGNI P, MANNELLA P, GAZZANIGA GM, PAOLINO RM, IACCARINO V, FELTRIN GP, ROVERSI R, MALLARNI G (1985). Percutaneous biliary drainage in neoplastic jaundice. Statistical data from a computerized multicenter investigation. *Acta Radiologica Diagnostica* **26**: 681–8.

REIMAN TH, BALFE DM, WEYMAN PJ (1987). Suprapancreatic biliary obstruction: CT evaluation. *Radiology* **163**: 49–56.

CHAPTER 3

Biliary Endoprosthesis

Andy Adam

Indications 33

Stent design 33

Insertion techniques 36

Complications and how to deal with them 41

Management of occluded stents 43

Clinical results 48

Future developments 50

References 50

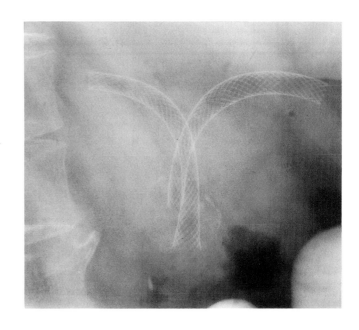

INDICATIONS

The main indication for the use of biliary endoprostheses is obstructive jaundice due to malignancy. Occasionally indwelling stents, especially ones inserted endoscopically, are employed for the temporary relief of biliary obstruction due to stones or benign strictures of the common bile duct (see Chapter 6) but as endoprostheses are usually employed for palliation rather than the elimination of the cause of biliary obstruction it is best to reserve them for patients with irresectable malignant tumours.

Some interventional radiologists believe that endoprostheses should be reserved for patients whose life expectancy is relatively short in order to avoid the problem of stent occlusion (Ferrucci, 1985). We believe that endoprostheses should be inserted in virtually all patients with malignant obstructive jaundice in whom their insertion is technically feasible. The only exceptions are patients who have preoperative biliary drainage in whom it is best to use a catheter providing internal/external drainage (see Chapter 2). Our preference for endoprostheses over internal/external catheters is based on a number of factors: external catheters are associated with complications such as occasional leakage of bile around the catheter, and infection and pain at the skin entry site. The latter two are particularly likely to occur if the catheter is left in place for longer than a few days. However, our main objection to long-term external catheters is the psychological discomfort inflicted on the patient: as most patients with malignant obstructive jaundice know that they have a terminal condition a permanent external catheter is a constant and unkind reminder of their short life expectancy.

The main objection to the use of indwelling stents has been the fact that they are likely to become occluded in patients with a relatively long life expectancy. However stent replacement is now a routine procedure: occluded stents projecting into the duodenum can be changed endoscopically with relative ease and various percutaneous methods are now available for changing stents the lower end of which lie above the ampulla of Vater.

STENT DESIGN

Plastic endoprostheses

Over the years many different types of plastic stents have been tried in an effort to combat the two main long-term complications of biliary endoprostheses, namely, migration and occlusion. There is an inverse relationship between the size of the lumen of biliary stents and the incidence of occlusion due to encrustation of bile; this has lead to a gradual increase in the size of plastic endoprostheses. However, large plastic stents require a large transhepatic track for their introduction which is associated with a greater incidence of haemobilia. In addition, the greater stiffness of the larger calibre introducing systems predisposes to buckling of wires at points of angulation in the track and possible perforation of bile ducts especially around the hilum of the liver. In view of these considerations most hepatobiliary radiologists consider that 12 Fr gauge is the optimum size for plastic endoprostheses.

Another factor which influences the incidence of bile encrustation is the material used for the construction of biliary endoprostheses. Research conducted *in vitro* and *in vivo* has shown that Teflon is associated with the lowest incidence of bile encrustation whereas polyurethane and polyethylene are associated with progressively higher rates of occlusion (Lammer *et al.*, 1986). However, in practice, the differences are not such as to override other considerations in the choice of stent.

Most plastic endoprostheses have side holes along their length. These probably do not contribute to better drainage of bile but increase the margin of error in positioning the endoprostheses should slippage of the stent position occur. However, there is evidence that the presence of side holes in biliary stents increases the rate of occlusion (Coene *et al.*, 1990) and it is probably best to use endoprostheses with end holes only, except in hilar strictures where the presence of side holes may be important in maintaining drainage of side branches of the duct containing the endoprosthesis.

In an effort to combat the problem of stent migration various designs have been introduced including spirals (Yeung *et al.*, 1988), mushroom tips at the proximal and distal ends of the stent, and subcutaneous plastic buttons attached to the endoprosthesis (Fig. 3.1). These have proved very successful in stabilizing biliary endoprostheses and migration is no longer a significant problem. Straight stents without protuberances can only be stabilized if they are of sufficient length (Coons *et al.*,

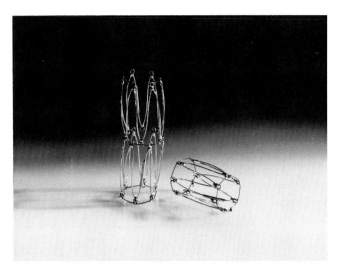

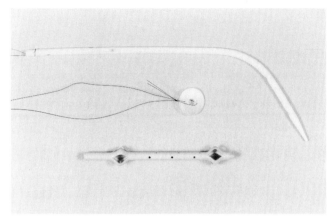

Fig. 3.2
The Carey–Coons and Miller biliary endoprostheses. The former is attached with silk sutures to a plastic button which can be buried subcutaneously to prevent migration of the stent.

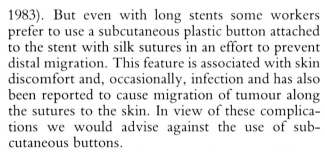

Fig. 3.1
The Rösch modification of the Gianturco endoprosthesis. Single and double stents are shown.

1983). But even with long stents some workers prefer to use a subcutaneous plastic button attached to the stent with silk sutures in an effort to prevent distal migration. This feature is associated with skin discomfort and, occasionally, infection and has also been reported to cause migration of tumour along the sutures to the skin. In view of these complications we would advise against the use of subcutaneous buttons.

A disadvantage of using length as a stabilizing factor is that the endoprosthesis has to be left protruding into the duodenum. Of course this is often inevitable even with short stents if the occluding lesion is in the lower common bile duct. However, we believe that in those patients with hilar lesions in whom it is possible to leave the lower end of the stent above the ampulla of Vater it is desirable to do so. Contamination with bacteria is probably a predisposing factor to stent occlusion by encrusted bile and is more likely to take place if the stent is in communication with duodenal contents. In addition, food material has been shown to occlude stents protruding into the duodenum. In most occluded endoscopically inserted stents it is the distal 20% of the endoprosthesis which projects into the duodenum which is the site of occlusion. The Carey–Coons endoprosthesis, made of Percuflex (Fig. 3.2), is very easy to introduce and even if used without the anchoring plastic button has reasonable stability because of its long length. However, the lower end must be left in the duodenum with the associated disadvantages described above. In addition, because Percuflex is soft, it may buckle when going around acute angles in the duct system. In our

opinion, the Miller double mushroom-tipped stent (Fig. 3.2) is one of the best plastic endoprostheses currently available as it combines excellent stability with ease of insertion and a low occlusion rate. The main disadvantage of this stent compared with some other stent designs is that it is slightly more difficult to replace percutaneously (see below).

Metallic stents

In recent years various types of metallic endoprostheses have become available commercially. These fall into two broad categories:

1 Stents which are inserted over an angioplasty balloon and expanded when in an optimum position across the biliary stricture (e.g. Strecker stent, Palmaz stent).
2 Self-expandable endoprostheses (e.g. Wallstent, Gianturco stent).

The main advantage of metallic stents is that they can be introduced in their contracted state through a very small calibre track and achieve a large internal lumen following expansion. This reduces the incidence of bile encrustation but it is difficult to be certain about the exact significance of this factor on the basis of currently available data. Metallic stents in their expanded state are partially embedded in the duct wall and this virtually eliminates the problem of migration. In addition, some stent designs incorporate sharp ends or barbs which also increase the security of maintaining stent position.

BALLOON-EXPANDABLE METALLIC STENTS

Balloon-expandable endoprostheses have to be inserted whilst positioned over the angioplasty catheter and the rigidity inherent in this design feature may make it difficult to advance the stent around acute angles in the biliary system and indeed can result in the endoprosthesis being displaced from its position over the balloon. However, sufficient dilation of the track can usually overcome this problem. A more serious problem associated with certain types of balloon-expandable stents is that it may be very difficult to achieve adequate overlap of short endoprostheses around sharp angles. This may result in ingrowth of tumour through the gaps between the stents.

SELF-EXPANDABLE METALLIC STENTS

The two main designs of self-expandable biliary endoprostheses are the Gianturco zig-zag stent (Cook Inc., Bloomington, IN) and the Wallstent (Medinvent SA, Lausanne, Switzerland). The Rösch modification of the Gianturco stent (Fig. 3.1), which is the type usually employed in the biliary tree, is constructed from a zig-zag of 0.018 inch stainless steel wires soldered together with a lead-free solder to form a closed ring (Martin *et al.*, 1990). A retaining suture through eyelets at each end control the diameter of the resultant cylinder. Most stents are used in tandem with a retaining suture woven through the eyelets of each stent. Paired 10 mm stents are usually employed for the common bile duct and at the hilum, and 8 mm stents in the intrahepatic biliary radicles. Placement is planned so that the critical stenosis is bridged by a double stent. The unmodified Gianturco stent is similar in design but is fitted near the midpoint with small barbs that engage in short strictures in order to prevent migration (Irving *et al.*, 1989). The stent is introduced through a 12–14 Fr introducer sheath. Gianturco stents are excellent when dealing with benign biliary strictures (see Chapter 4). However, they are unsuitable for malignant lesions because ingrowth of tumour results in their rapid occlusion in many cases.

The Wallstent biliary endoprosthesis is identical in design to endovascular Wallstents: it consists of surgical grade stainless steel alloy filaments woven in a tubular fashion (Fig. 3.3). Since the cross-points of the filaments are not soldered together, the stent is pliable, self-expanding and flexible in the longitudinal axis (Zollikofer *et al.*, 1988). It is so elastic and pliable that its diameter can be substantially reduced by moderate elongation, which allows it to

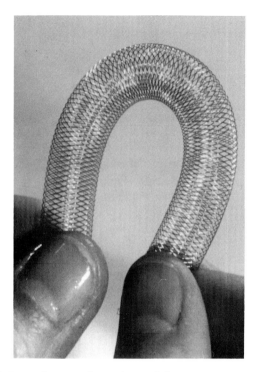

Fig. 3.3 *Wallstent endoprosthesis (fully expanded)*

be mounted on a special catheter for introduction into the biliary system. The mounted stent is constrained on the delivery catheter by a doubled over membrane (Fig. 3.4). To release the stent, the outer membrane is slowly withdrawn, which allows the stent to return to its original diameter by its own elastic expansile force. Two radio-opaque markers on the delivery catheter allow the proximal and distal ends of the prosthesis to be identified at the time of release, hence enabling exact placement.

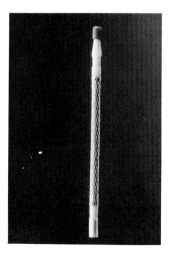

Fig. 3.4 *Wallstent endoprosthesis mounted on a delivery catheter*

INSERTION TECHNIQUES

All biliary endoprostheses except the Wallstent (see below) are best inserted in a two-step procedure. Initially percutaneous biliary drainage (PBD) is performed as described in Chapter 2. The endoprosthesis is inserted 4–7 days after the PBD procedure. This time interval makes endoprosthesis insertion much easier, faster and less painful for the patient, and is associated with a lower complication rate than a single-step approach.

The items of equipment needed for endoprosthesis insertion are a stiff-bodied but soft-tipped guidewire, biliary dilators, a pusher catheter and a suture. The patient is prepared as for the initial biliary drainage procedure. The soft-tipped exchange guidewire is placed through the existing transhepatic catheter and positioned with the soft tip at or just beyond the ligament of Treitz. If any difficulty is experienced at directing the wire into the fourth part of the duodenum a femorovisceral (cobra) catheter will facilitate this manoeuvre. Following removal of the transhepatic catheter, biliary dilators of the appropriate size are passed over the wire. A suture is looped through a proximal side hole of the endoprosthesis. The endoprosthesis is then placed coaxially over the wire or the 'rider' catheter, depending on design. The stent is then pushed into the desired position by a plastic pusher catheter (Fig. 3.5a). The location of the stent is

assessed fluoroscopically by injection of contrast medium through the pusher catheter. If a more proximal position is required, the endoprosthesis is adjusted by using the suture passed through the proximal side hole. After the optimum position has been obtained, the suture is removed by gentle traction while the pusher catheter prevents proximal migration. A small calibre catheter is then advanced over the guidewire until its tip lies just into the proximal end hole of the endoprosthesis and the guidewire is removed. The external catheter remains in place for approximately 24 hours and allows repeat cholangiography on the following day to establish that the endoprosthesis remains in a satisfactory position. After completion of the angiogram the external catheter is removed (Fig. 3.5b).

Certain modifications of this technique are required for various specific catheter designs. The double-mushroom (Miller) endoprosthesis cannot be pushed safely through the liver parenchyma and requires placement of a Teflon sheath of sufficient lumen to accommodate the endoprosthesis (Fig. 3.6).

Antibiotics should continue to be administered for at least 48 hours following removal of the external catheter in order to prevent cholangitis. It is possible that continuous administration of antibiotics may minimize bacterial colonization of the endoprosthesis and prolong patency.

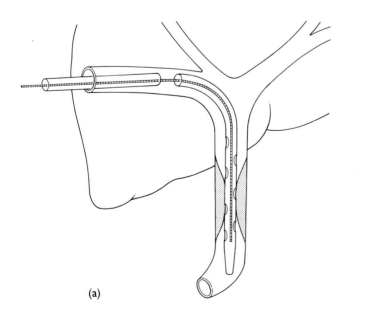
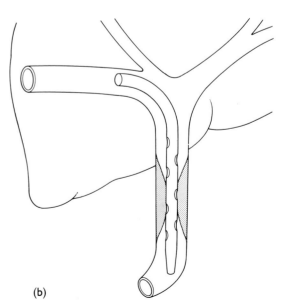

(a)   (b)

Fig. 3.5 *Biliary endoprosthesis insertion*
(a) The stent is being advanced into position using a plastic pusher over a guidewire. (b) Following withdrawal of the wire and pusher the endoprosthesis alone is left in position.

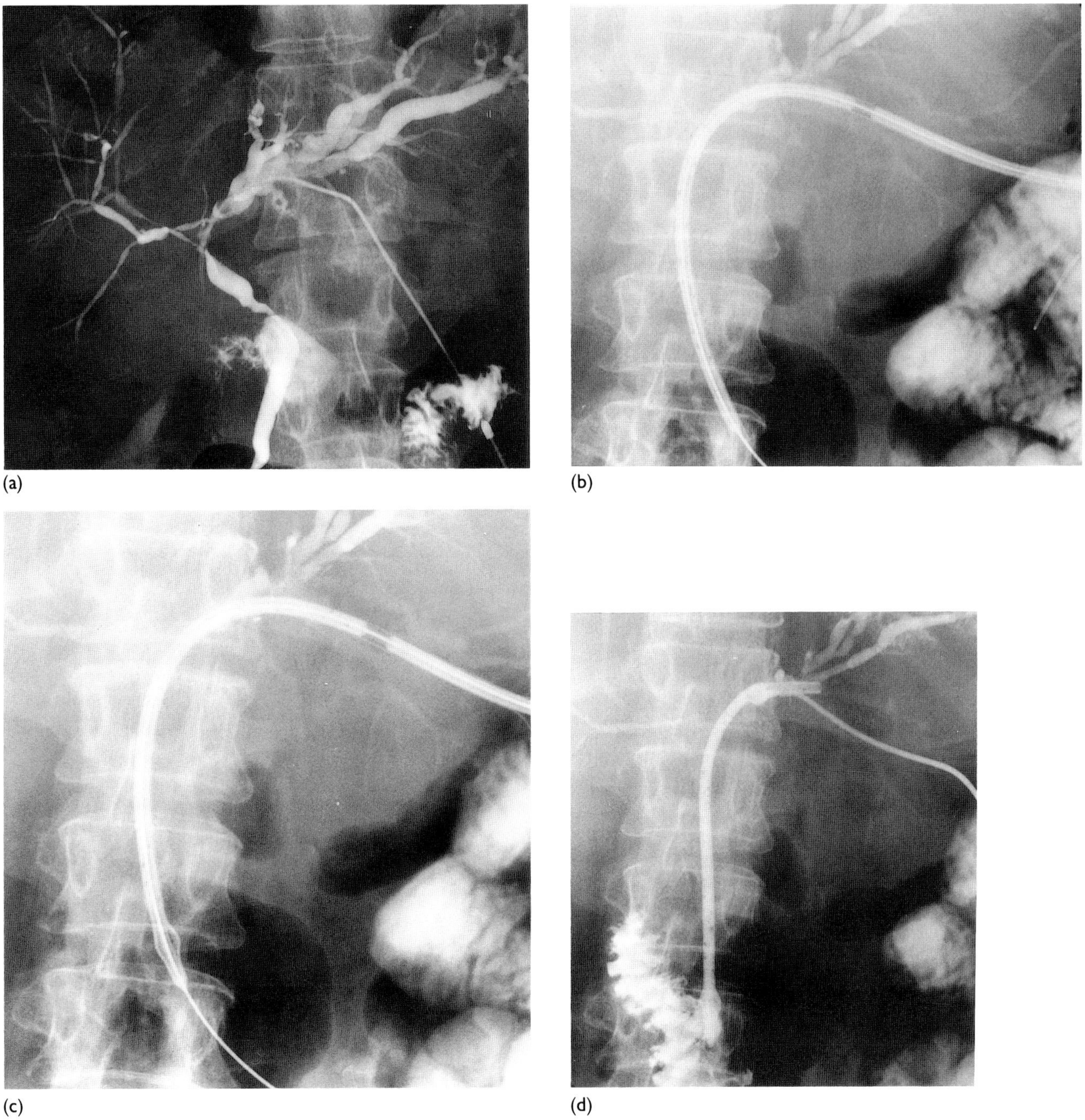

(a)

(b)

(c)

(d)

Fig. 3.6 *Miller biliary endoprosthesis insertion*
(a) Injection of contrast medium via an external catheter into the left ducts shows a stricture of the left and right hepatic ducts and the upper common hepatic duct. (b) The endoprosthesis is being pushed into position. This is done through a sheath (not visible on the radiograph).
(c) Partial withdrawal of the sheath has resulted in expansion of the distal mushroom tip. (d) The stent is completely released. Injection of contrast medium through a temporary external catheter shows good flow through into the duodenum.

Placement of Wallstent endoprostheses

The principles of placement of Wallstent endoprostheses are similar to those of plastic stent insertion. However, there are certain important differences of detail and these are outlined below.

The essential steps are shown in Fig. 3.7: the stricture is dilated with an 8–10 mm angioplasty balloon. An endoprosthesis of appropriate length is then chosen. For strictures at the hilum of the liver or the mid-common duct we prefer to use a long endoprosthesis which will have its proximal end peripherally in an

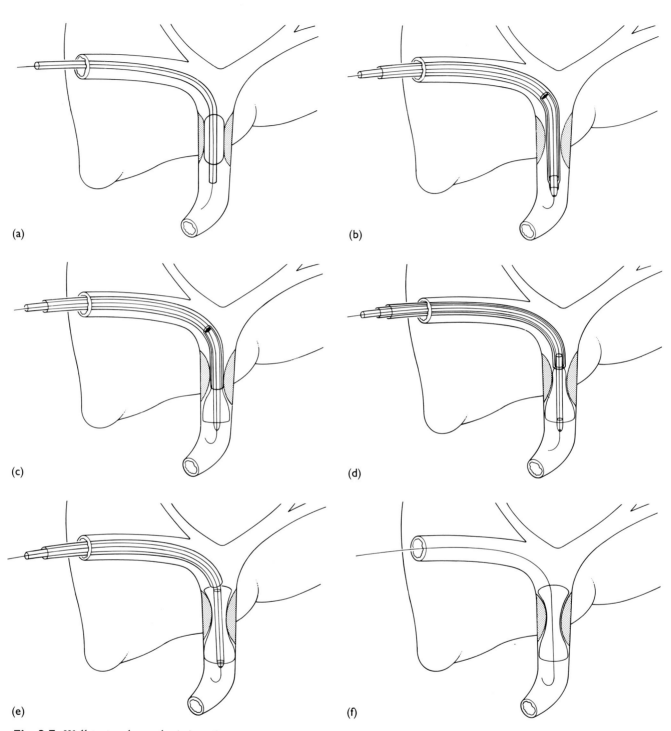

Fig. 3.7 *Wallstent endoprosthesis insertion*
(a) The stricture is dilated with a low-profile, high-pressure balloon. (b) The stent in its contracted state is positioned across the stricture. (c) Partial withdrawal of the plastic rolling membrane allows the stent to expand gradually from its distal end. (d) Immediately before the stent is completely released its position is finally adjusted. (e) Complete withdrawal of the membrane releases the stent. (f) The released stent usually remains partially contracted. (g) The stent is dilated with a balloon. (h) The wire and balloon catheter are withdrawn leaving only the stent in position.

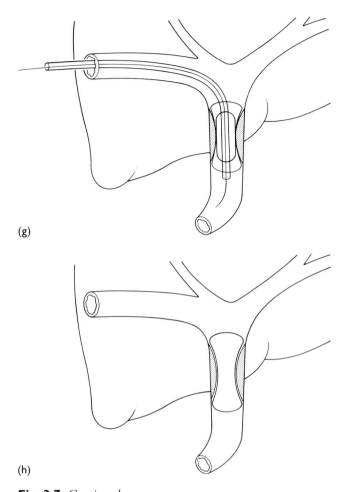

(g)

(h)

Fig. 3.7 *Continued*

intrahepatic duct and its lower end 1–2 cm above the ampulla of Vater. In cases of low common bile duct lesions the stent obviously must project through the ampulla into the duodenum. The policy of using long stents does not lead to obstruction of side branches draining into or converging with the duct containing the endoprosthesis within the liver. Such branches continue to drain through the side wall of the Wallstent even months after the initial placement of the endoprosthesis. The use of long stents delays occlusion caused by overgrowth of tumour above or below the stent. In patients with pancreatic carcinoma occluding the lower common bile duct a long stent projecting above the hilum of the liver will help to prevent hilar duct obstruction due to enlarged lymph nodes containing metastases. Following selection of a stent of appropriate length, the introducing catheter is advanced until the distal metallic marker which corresponds with the lower end of the stent is immediately above the ampulla of Vater (except in patients with lower common bile duct lesions in whom the stent projects into the

duodenum). The plastic membrane is then withdrawn until approximately half the endoprosthesis is expanded. The position of the upper end of the stent is then adjusted by withdrawing the endoprosthesis to an appropriate point within a peripheral hepatic duct. The stent is then released by complete withdrawal of the plastic membrane. The 10 mm angioplasty balloon catheter is used to expand the stent at the point where it crosses the malignant stricture until it achieves its maximum internal diameter of 1 cm. However, this manoeuvre is not essential; the stent exerts a continuous radial force and will expand spontaneously over the following few days. Following release of the endoprosthesis a 6 Fr temporary external drainage catheter is advanced through the stent over the guidewire and is left in place overnight for check cholangiography to be carried out on the following day, after which the catheter is removed (Fig. 3.8). It is wise to preload an 8 Fr plastic sheath over the introducing catheter of the Wallstent. The sheath is kept over the part of the catheter which is outside the body and is not usually employed at all during insertion of the stent. However, if the endoprosthesis is pulled proximally during the procedure, but has not yet been completely released the sheath makes it possible to reposition it: the sheath is advanced over the Wallstent thus closing it. The endoprosthesis, enclosed within the sheath, is then advanced to the desired position. The sheath is then withdrawn and the stent is released.

The small calibre introducing catheter of the Wallstent endoprosthesis makes it possible to insert such stents in a single-step procedure in a significant number of patients. This has several advantages:

1 It foreshortens the patient's stay in hospital.
2 It can be performed under sterile conditions without external bacterial contamination of the bile ducts.
3 The patient is spared the discomfort of a second procedure and the risk of dislocation of the external drainage catheter.

If it is desired to drain both sides of the liver in the case of a hilar malignant lesion, this can be done by deploying two Wallstents side by side, either through the same transhepatic track or through two different tracks, one on the left and one on the right side. If a single transhepatic track is to be used a Y arrangement of stents may be employed (Fig. 3.9). In the case of right-sided punctures one stent is deployed extending from the right hepatic duct to the left hepatic duct and another extending from the right hepatic duct to the common bile duct. The duct containing the two stents will expand to accommo-

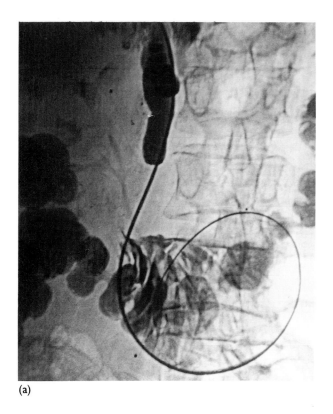

(a)

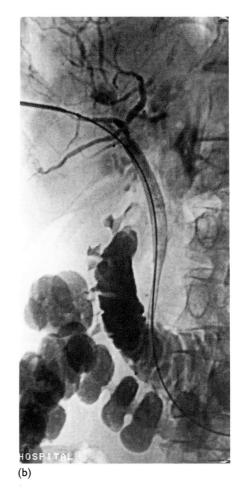

(b)

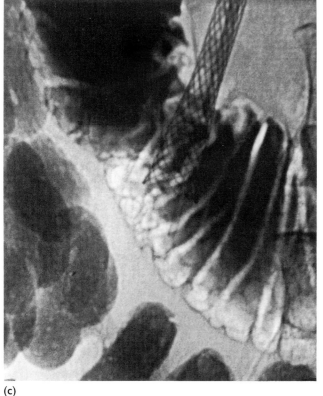

(c)

Fig. 3.8
There is complete obstruction of the lower end of the common bile duct (a) A catheter and guide wire have been advanced into the duodenum. (b) A Wallstent endoprosthesis has been inserted extending from the hilum of the liver to the duodenum. (c) The lower end of the stent lies within the duodenum. A temporary external catheter is seen within the endoprosthesis. This was removed following cholangiography on the following day.

date them without difficulty. In a modification of this method the stents are arranged in a T configuration (LaBerge *et al.*, 1990). The initial stent is placed transhepatically into the common bile duct, and the second stent (forming the T) is placed transductally from the entry duct extending to the contralateral duct. The initial stent is released in the standard fashion from the duct of entry into the common bile duct (this will be referred to as the ipsilateral stent). After placement of the ipsilateral stent, a Terumo Glide wire is advanced through the wire mesh of the stent into hepatic confluence and then manipulated into the contralateral duct. A high-pressure 8 mm balloon is advanced over the wire and through the mesh. Inflation of the balloon dilates both the wire mesh opening and the contralateral hilar lesion. A stent is then deployed in a transductal position.

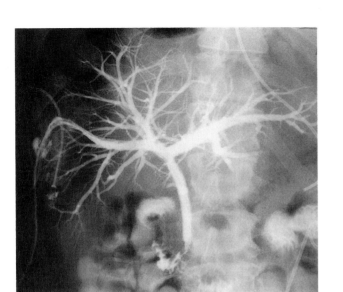

(a)

Fig. 3.9 *Radiographs of a 76-year-old man with cholangiocarcinoma*
(a) Contrast medium flows through stents inserted from the same puncture on the right side. (b) One stent extends from right to left and the other from the right duct to the common bile duct. The upper ends of the stents have not expanded fully because of lack of space in the right hepatic duct. Reproduced from Adam *et al.* (1991) with permission.

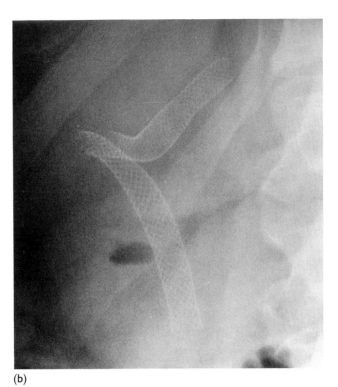

(b)

Subsequently, a Glide wire is manipulated through the wire mesh of the transductal stent down into the ipsilateral stent and into the common bile duct. The 8 mm balloon is then used to create an opening in the transductal stent. This arrangement has the theoretical advantage of not overexpanding the common bile duct by the presence of two stents side by side.

If two separate punctures are to be used the stents extend from the right and left hepatic ducts to lie side by side in the common bile duct (Fig. 3.10). When Wallstents are deployed from two separate punctures it is best to release them simultaneously (using two operators) to avoid displacement of the first stent during the subsequent release of the second one. A different, and in our view a preferable, method of achieving the same result involves the use of two long 8 Fr sheaths. These are deployed in the ducts in which the stents are to be placed, for example, one extending from the left hepatic duct to the lower common bile duct and one from the right hepatic duct to the lower common bile duct. The endoprostheses are then partially released *within* the sheaths until most of each stent has been expanded but not detached from the introducing catheter. The sheaths are then withdrawn before the endoprostheses are released completely. Although the T and Y arrangements are technically elegant, our own preference is for the side-by-side deployment of stents through two separate punctures. The side-by-side configuration allows a more proximal placement of the upper end of the endoprostheses within intrahepatic ducts thus delaying occlusion by tumour overgrowth. In addition, the procedure of stent overlap for the purpose of unblocking a stent occluded by overgrowth of tumour is technically more straightforward when a side-by-side arrangement is used. Also, this arrangement places more barriers to the growth of hilar tumours and is more likely to result in at least one of the two endoprostheses remaining patent for a prolonged period of time.

COMPLICATIONS AND HOW TO DEAL WITH THEM

The main complications of biliary endoprostheses are migration, proximally or distally to the obstructing lesion, overgrowth of the proximal or distal end of the endoprosthesis by the underlying malignant process, occlusion of the lumen of the endoprosthesis, or very rarely perforation of the duodenum. Metallic stents are associated with the additional complication of ingrowth of tumour through their side walls.

Duodenal perforation is uncommon and its inci-

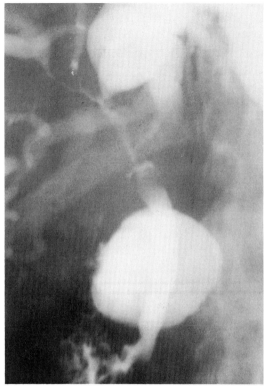

(a)

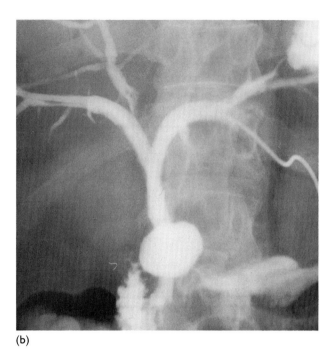

(b)

Fig. 3.10 *Radiographs of a 58-year-old woman with cholangiocarcinoma*
(a) Strictures of right and left hepatic ducts and common hepatic duct. (b) Contrast medium flows through stents inserted via separate punctures. (c) The stents are shown unobscured by contrast medium. Reproduced from Adam *et al.* (1991) with permission.

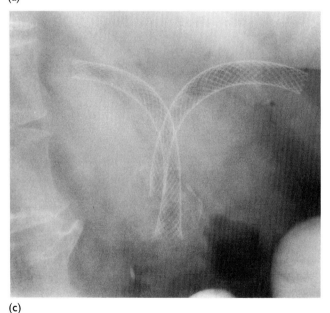

(c)

Neumayer, 1986).

Dislocation of an endoprosthesis usually occurs within 24 hours of its insertion but may be delayed. In those cases in which a plastic stent migrates while a temporary external drainage catheter is still *in situ* it is easy to use the catheter to correct the position of the stent: a guidewire is advanced through the drainage catheter into the proximal end hole of the migrated stent. The catheter is then exchanged for a small, low-profile angioplasty balloon catheter such an Olbert 4.8 Fr catheter carrying a 4 mm balloon (Meadox Inc.). The balloon is inflated into the proximal part of the endoprosthesis and the catheter is then withdrawn until the stent is in a satisfactory position (Harries-Jones *et al.*, 1982). The balloon is then deflated and the catheter withdrawn. A similar technique can, of course, be used for late dislocations but in those circumstances a new PBD procedure has to be performed to gain access to the proximal end of the migrated stent.

If the dislocated endoprosthesis cannot be re-catheterized it is often possible simply to push the nonfunctioning stent into the duodenum before replacing it with a new one. Alternatively, a trans-duodenal approach can be used if a skillful endoscopist is available. If the endoprosthesis can be

dence can be minimized by ensuring that any part of the plastic stent which projects into the duodenum lies parallel to the bowel rather than perpendicular to it. Wallstent endoprostheses have sharp ends which should not be in contact with the distal wall of the duodenum opposite the ampulla of Vater. Although duodenal perforation can be a very serious complication, it has been treated by simple endoprosthesis repositioning with no clinical sequelae (Lammer and firmly grasped by the endoscopist, it is possible to pull it into the duodenum. In some cases it will be possible for the endoscopist to place a new stent of sufficient calibre through the ampulla of Vater.

Obstruction of plastic biliary endoprostheses occurs in 20–30% of patients who live longer than 6–18 months (Teplick, 1991). This complication results in a nonfunctioning endoprosthesis which manifests as recurrent jaundice. If the bile is infected, jaundice may be accompanied by cholangitis.

All endoprostheses that are currently in use will eventually obstruct. It is not possible to predict how long a given endoprosthesis will remain patent and the functional period may be as little as 3–4 weeks or as long as several years (Teplick, 1991). The average patency is less than 1 year, but most patients with advanced malignancies will succumb to their disease before the endoprosthesis occludes.

The mechanism of reocclusion of plastic and metal endoprostheses is different. In plastic endoprostheses, bacterial contamination causes decomposition of the bile and subsequent encrustation. In metal endoprostheses the usual cause of occlusion is overgrowth of tumour above or below the stent or, ingrowth of the neoplasm between the struts of the endoprosthesis (Lammer, 1990).

MANAGEMENT OF OCCLUDED STENTS

Plastic stents

It is possible to insert a Chiba needle into a side hole of an occluded plastic stent and inject saline forcibly to reopen the endoprosthesis without the need for transhepatic catheter placement (Gibson, 1986). However, it is likely that the bacteria which initiate bile encrustation in the first place will lead to relatively rapid reocclusion. It is best to remove occluded endoprostheses and replace them with new ones.

Various methods of stent replacement have been described. Our preferred method for changing an occluded endoprosthesis which does not have any protuberances (e.g. the Carey–Coons endoprosthesis) is to use a Chiba needle to catheterize the proximal end hole of the stent under biplane fluoroscopic guidance. Then a coaxial catheter introduction set (e.g. Neff) is employed to insert a 0.038 inch guidewire into the endoprosthesis. Following this, a low-profile angioplasty catheter is advanced into the

stent and the balloon inflated just inside its proximal end. The occluded endoprosthesis can then be withdrawn through the skin or pushed into the duodenum (Adam, 1987). As the wire is already through the stricture it is easy to advance a new endoprosthesis to the appropriate position (Figs 3.11 and 3.12).

It is too uncomfortable to remove double mushroom-tipped stents through the skin and it is best to push them into the bowel to be retrieved endoscopically or to be passed with the patient's faeces. Our method for doing this is shown in Figs 3.13 and 3.14. The biliary tree is punctured and opacified with contrast medium. A guidewire and catheter are inserted and negotiated alongside the stent and through the stricture until their tips are in the bile duct distal to the stricture. The catheter/guidewire combination is then manipulated into the distal mushroom and the guidewire advanced through the hole in the distal mushroom tip. The catheter is then advanced, pushing the stent distally. The stent can usually be pushed in this manner so that its distal tip rests in the duodenum and its proximal end is in or past the bile duct structure. Further advancement of the stent past this position is often not possible because the catheter/wire tends to loop in the duodenum. If this is the case the catheter and wire are withdrawn from the distal mushroom and pulled back up the bile duct. The proximal lumen of the stent is then cannulated, either via the proximal end hole or mushroom, and the catheter is readvanced pushing the entire stent into the duodenum. A sharp tug on the catheter is usually necessary to withdraw it from the stent, leaving the endoprosthesis entirely within the duodenum. A replacement endoprosthesis can then be inserted at the same session (Yeung et al., 1989).

Metallic stents

The usual cause of occlusion of a Wallstent endoprosthesis is overgrowth of tumour above or below the stent. In order to deal with this problem percutaneous cholangiography is first performed to demonstrate the dilated ducts above the obstructed endoprosthesis. Then a catheter is inserted in the biliary tree through as peripheral a puncture as possible. The catheter is advanced to the point where the tumour is causing occlusion of the stent. Injection of contrast medium through this catheter may show a track through the tumour and enable the catheter or a guidewire to be manipulated through the obstruction. If no such track is demonstrated a Glide wire (Terumo Inc.) will usually cross the obstruction and can then be advanced into the duo-

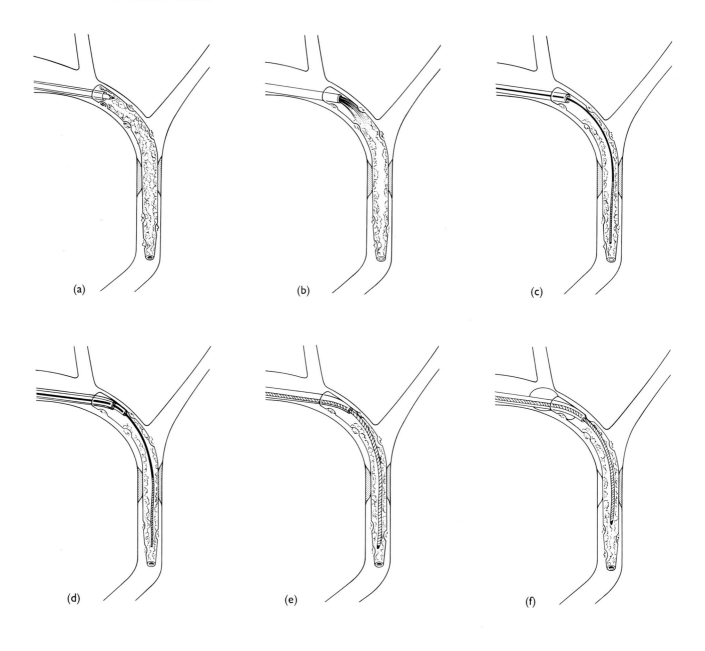

(a) (b) (c)

(d) (e) (f)

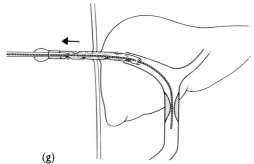

(g)

Fig. 3.11 *Stages of removal of occluded Carey–Coons biliary endoprosthesis using a coaxial catheter introduction set*
(a) The proximal end of the stent is cannulated with a 21 gauge Chiba needle. (b) The stent is washed out with saline and contrast medium. (c) The 0.018 inch (0.46 mm) wire is advanced down the length of the endoprosthesis. (d) The 6 Fr sheath with its dilator are advanced over the wire. (e) The 0.018 inch wire and the dilator have been exchanged for a 0.038 inch (0.97 mm) wire. (f) The sheath is replaced with a catheter carrying a 4 mm balloon which is inflated within the stent but partially projects from its proximal end to provide a smooth shoulder for withdrawal. (g) The stent is removed together with the balloon catheter.

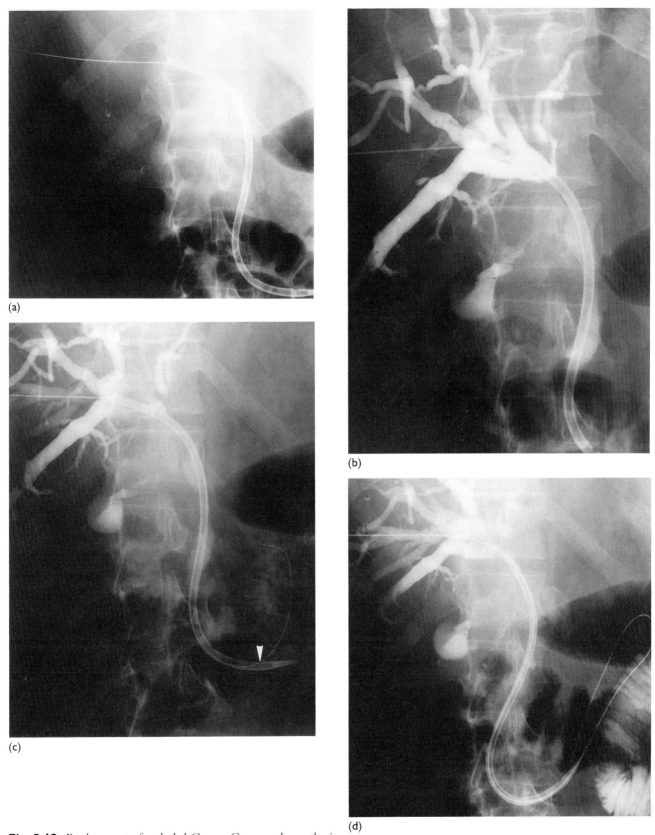

Fig. 3.12 *Replacement of occluded Carey–Coons endoprosthesis*
(a) Cannulation of the upper end hole of the occluded stent with a Chiba needle. (b) Injection of contrast medium demonstrates no flow of contrast medium through the stent. A little contrast medium flows around the endoprosthesis. (c) The 0.018 inch wire has been advanced through the stent and exists via one of the lower side holes (arrowhead). (d) The 0.038 inch wire is now within the stent and exists via the lower end hole. (e) The balloon catheter is within the endoprosthesis. The balloon is not visible but its position is indicated by two metal markers within the endoprosthesis. (f) The occluded Carey–Coons stent has been replaced with a Miller endoprosthesis. A temporary external catheter is *in situ*.

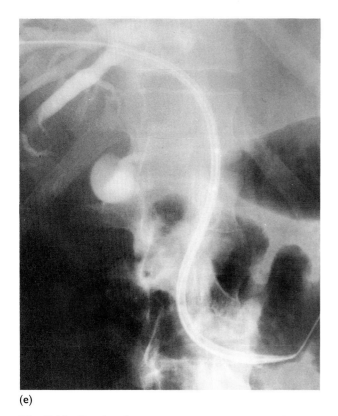

(e)

Fig. 3.12 *Continued*

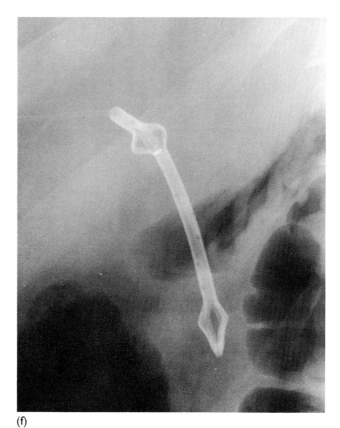

(f)

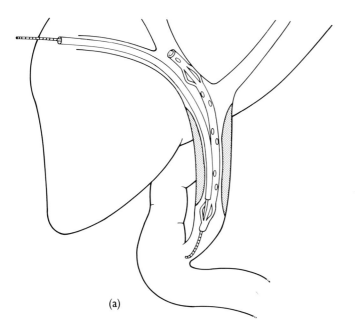

(a)

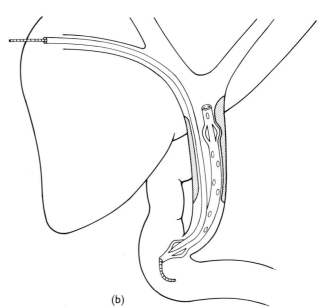

(b)

Fig. 3.13 *Miller endoprosthesis removal*
(a) The catheter/guidewire combination is advanced through the stricture alongside the stent, the distal mushroom entered and the guidewire advanced through the distal end hole. (b) The catheter/guidewire combination is advanced, pushing the lower end of the stent through the ampulla. (c) Once the distal mushroom has entered the duodenum further attempts at advancement of the stent often result in looping of the catheter and guidewire in a duodenum and failure of the distal mushroom to move any further. (d) The proximal end hole of the stent has been cannulated and the stent is advanced into the duodenum.

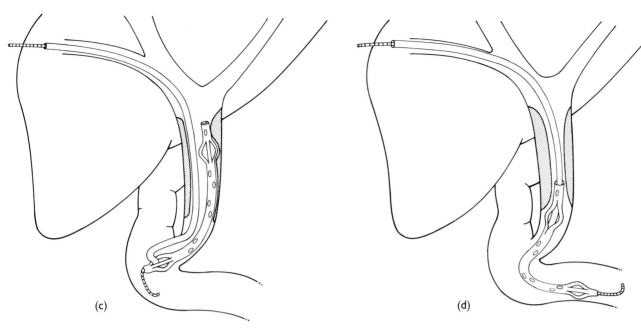

(c)   (d)

Fig 3.13 *Continued*

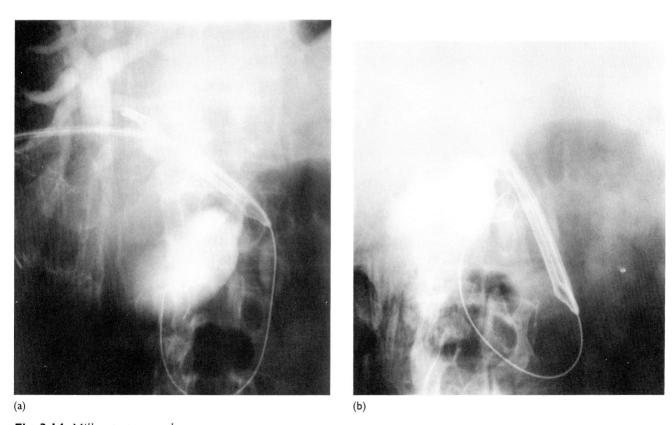

(a)   (b)

Fig. 3.14 *Miller stent removal*
(a) A catheter/guidewire combination has been advanced through the distal mushroom. The wire has entered the duodenum. (b) The distal mushroom has been pushed into the duodenum. (c) The catheter has been manipulated into the proximal mushroom. The stent has been pushed into the third part of the duodenum. (d) The stent is now in the fourth part of the duodenum.

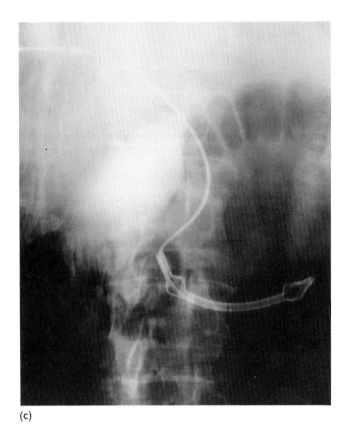

(c)

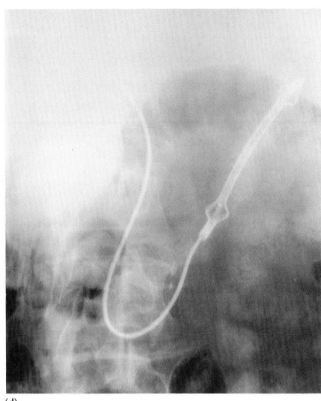

(d)

Fig 3.14 *Continued*

denum. The catheter is advanced over the Glide wire which is then exchanged for a super stiff wire (Boston Scientific A/S, Denmark). Following removal of the initial catheter the introducing catheter of the Wallstent is used to overlap a second stent with the first one. The ends of the endoprosthesis should project well clear of the occluded tumour. This manoeuvre restores patency of the biliary system and can be used on more than one occasion if recurrent occlusion occurs (Jackson *et al.*, 1991) (Fig. 3.15).

In cases of advanced hilar malignancy extending into the liver it may be necessary not only to unblock Wallstent endoprostheses using the method described above but also to drain the contralateral lobe if that has not already been stented. This will necessitate the advancement of a guidewire thorough the mesh of the original stent. This manoeuvre is best carried out using a Glide wire (Terumo Inc.). A low-profile angioplasty catheter is then advanced over the wire and the balloon used to split the struts of the endoprosthesis and create a hole in the side of the stent. A second endoprosthesis can be placed through this hole to establish drainage of the lobe which was originally left undrained.

CLINICAL RESULTS

A very large number of factors influences the outcome of patients in whom biliary endoprostheses have been inserted (Mueller *et al.*, 1985). In addition, the follow-up of patients in some series is incomplete. Comparison of the results of different studies is therefore very difficult indeed and a detailed analysis of a large number of series would be extremely complex and of doubtful practical application. We prefer, therefore, to present our current practice and to indicate, where appropriate, how this has been influenced by our experience and by the collective results published to date.

We have reviewed our experience with 98 consecutive patients with malignant obstructive jaundice treated with plastic endoprostheses and 142 consecutive patients managed with Wallstents during a 48 month period. The overall median patient survival and re-intervention rates to deal with stent occlusion rates were 173 days and 43%, respectively, for the plastic stent patients. The figures for the Wallstent patients were 186 days and 12%, respectively. The five patients with Wallstent occlusion due to tumour overgrowth were treated

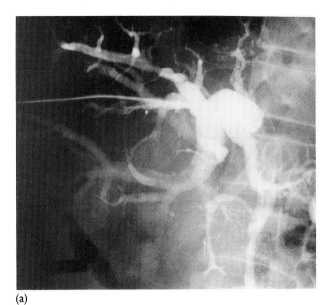

(a)

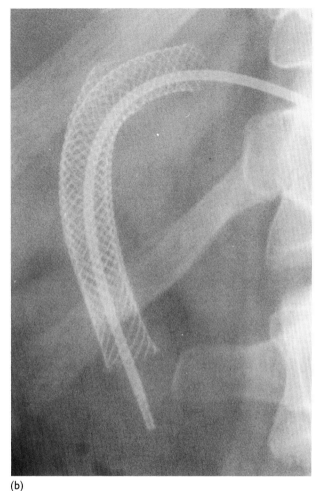

(b)

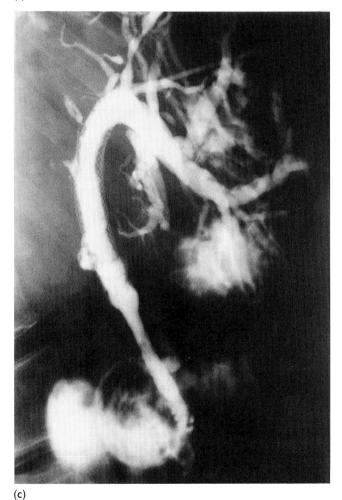

(c)

Fig. 3.15 *Management of occluded Wallstent endoprosthesis*
(a) Lateral view of percutaneous cholangiogram showing occlusion of a Wallstent extending from the left hepatic duct to the lower common bile duct. The occlusion was due to overgrowth of tumour above the endoprosthesis. (b) A second Wallstent has been inserted overlapping the first and projecting approximately 2 cm above the upper end of the first Wallstent. (c) Injection of contrast medium through a temporary external catheter following release of the second Wallstent shows free flow into the duodenum.

successfully by insertion of another stent overlapping the first as described above. The insertion of Wallstents was easier, more rapid and required less analgesia. Cholangitis requiring treatment with antibiotics was seen in 28% of the plastic stent patients and 6.5% of the Wallstent patients. Haemobilia requiring transfusion was seen in 5% of the plastic stent patients but in only 1.5% of the Wallstent patients.

The published rates of occlusion of plastic endoprostheses vary enormously, ranging from 1.5% (Coons and Carey, 1983) to 23% (Mueller *et al.*, 1985). We agree with Teplick (1991) that the true figure is approximately 20–30% in patients who live longer than 6–18 months. Despite this we prefer to use endoprostheses in virtually all patients with malignant obstructive jaundice rather than employ long-term internal/external catheter drainage. There are various solutions to the problem of endoprosthesis occlusion, some of which have been described in this chapter. Most patients, given the choice, prefer to have an additional procedure for an occluded stent to be changed, rather than to be left with a long-term external drainage catheter.

There are diverse opinions about relative indications for the use of metallic and plastic stents (Gordon *et al.*, 1990; Lammer, 1990). The use of inflexible stents such as the Gianturco and Palmaz endoprostheses is probably not indicated in malignant strictures, especially when dealing with masses at the hilum of the liver. This is because gaps between such stents used in tandem allow ingrowth of tumour with consequent recurrent jaundice. In addition, the large distance between the struts of the Gianturco stent is associated with a high rate of ingrowth of tumour (Irving *et al.*, 1989). It is possible that covering the Palmaz and Gianturco stents with plastic will prevent this complication but may be associated with a high incidence of occlusion due to bile encrustation.

Occlusion due to tumour ingrowth is infrequent with Wallstent endoprostheses and the rate of other complications is low (Adam *et al.*, 1991). In our opinion this is the best biliary stent currently available. It is easy to use and relatively atraumatic, thus allowing single session placement in a substantial proportion of patients. This reduces the total cost of endoprosthesis insertion when the number of days spent in hospital is taken into account.

FUTURE DEVELOPMENTS

Plastic endoprostheses are constantly being improved. It is now possible to bind antibiotics to drainage catheter materials (Trooskin *et al.*, 1985). By creating a local high concentration of antibiotic in the tube lumen, bacterial colonization, bile deconjugation and tube encrustation should be reduced, thus prolonging endoprosthesis patency. It may be possible to 'recharge' the tube by administration of antibiotics at intervals.

Metallic stents may also be revolutionized by the use of new materials such as Nitinol, a metal with 'memory' which allows a stent to adopt a predetermined shape when it reaches body temperature (Rabkin *et al.*, 1989).

In future it may be possible to use strong magnetic fields to induce electric currents in metallic endoprostheses in order to heat them to a predetermined temperature at regular intervals thus preventing tumour growth in the vicinity of the stent.

REFERENCES

ADAM A (1987). Use of the modified Cope introduction set for transhepatic removal of obstructed Carey-Coons biliary endoprosthesis. *Clinical Radiology* 38: 171–4.

ADAM A, CHETTY N, RODDIE M, YEUNG E, BENJAMIN IS (1991). Self-expandable stainless steel endoprostheses for the treatment of malignant bile duct obstruction. *American Journal of Roentgenology* 156: 321–5.

COENE PPLO, GROEN AK, CHENG J, OUT MMJ, TYTGAT GNJ, HUIBREGTSE K (1990). Clogging of biliary endoprostheses: a new perspective. *Gut* 31: 913–17.

COONS HG, CAREY PH (1983). Large-bore, long biliary endoprostheses (biliary stents) for improved drainage. *Radiology* 141: 799–801.

FERRUCCI JT (1985). Biliary endoprosthesis. In: *Interventional Radiology of the Abdomen*, pp. 267–81. Edited by Ferrucci, JT, Wittenberg J, Mueller PR, Simeone JF. W.B. Saunders, Philadelphia.

GIBSON RN (1986). Biliary endoprosthesis blockage clearance using a 22 gauge needle (technical note). *American Journal of Roentgenology* 147: 404–5.

GORDON RL, DICK BW, LABERGE JM, DOHERTY MM, RING EJ (1990). Clinical comparison of percutaneous use of metallic expandable Wallstents and conventional plastic endoprostheses in malignant biliary obstruction. *Radiology* 177 (P): 137.

HARRIES-JONES EP, FATAAR S, TUFT RJ (1982). Repositioning of biliary endoprosthesis with Grüntzig balloon catheters. *American Journal of Roentgenology* 138: 771–2.

IRVING JD, ADAM A, DICK R, DONDELINGER RF, LUNDERQUIST A, ROCHE A (1989). Gianturco expandable metallic biliary stents: Results of a European clinical trial. *Radiology* 172: 321–6.

JACKSON JE, RODDIE ME, CHETTY N, BENJAMIN IS, ADAM A (1991). The management of occluded metallic self-expandable biliary endoprostheses. *American Journal of Roentgenology* 157: 219–222.

LABERGE JM, DOHERTY M, GORDON RL, RING EJ (1990). Hilar malignancy: Treatment with an expandable metallic transhepatic biliary stent. *Radiology* 177: 793–7.

LAMMER J (1990). Biliary endoprostheses. Plastic versus metal stents. *Radiologic Clinics of North America* 28: 1211–22.

LAMMER J, NEUMAYER K (1986). Biliary drainage endoprostheses: Experience with 201 placements. *Radiology* 159: 625–9.

LAMMER J, STOEFFLER G, PETEK G et al. (1986). In-vitro long term perfusion of different materials for biliary endoprostheses. *Investigative Radiology* 21: 329.

MARTIN EC, LAFFEY KJ, BIXON R, GETRAJDMAN GI (1990). Gianturco-Rosch biliary stents: Preliminary experience. *Journal of Vascular and Interventional Radiology* **1**: 101–5.

MUELLER PR, FERRUCCI JI JR, TEPLICK ST (1985). Biliary stent endoprosthesis: Analysis of complications in 113 patients. *Radiology* **156**: 637–9.

RABKIN JK, TIMOSHIN AD, MEDNIK GI, MOROZOVA M, NELYUBIN SP (1989). Biliary stent using a Nitinol spiral – an experimental and clinical study. *Journal of Interventional Radiology* **4**: 135–9.

TEPLICK SK (1991). Percutaneous transhepatic insertion of biliary endoprostheses. In: *Current Practice of Interventional Radiology*, pp. 557–62. Edited by Kadir S. B.C. Decker Inc., Philadelphia.

TROOSKIN SZ, DONETZ AP, HARVEY RA, GRECO RS (1985). Prevention of catheter sepsis by antibiotic bonding. *Surgery* **97**: 547–51.

YEUNG E, O'DONNELL C, CARVALHO P, ADAM A (1989). A new technique for removal of double mushroom-tipped biliary endoprostheses. *American Journal of Roentgenology* **152**: 527–8.

YEUNG EYC, ADAM A, GIBSON RN, BENJAMIN IS, ALLISON DJ (1988). Spiral-shaped biliary endoprosthesis: Initial study. *Radiology* **168**: 365–9.

ZOLLIKOFER CL, LARGIADER I, BRUHLMANN WF, UHLSCHMID GK, MARTY AH (1988). Endovascular stenting of veins and grafts: Preliminary clinical experience. *Radiology* **167**: 707–12.

CHAPTER 4

Interventional radiology for benign biliary strictures

Robert N. Gibson and Andy Adam

Aetiology, location and presentation   53

Radiological assessment   53

Contraindications to intervention   54

Patient preparation   54

Anaesthesia   54

Imaging guidance   54

Radiological biliary access   55

Techniques for stricture dilation   60

Postdilation stenting   60

Metal stents   61

Follow-up   61

Management of associated calculi   64

Complications   65

Results   65

Role of radiological intervention in relation to surgical management   65

References   65

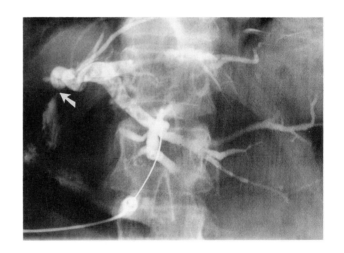

The management of patients with benign biliary strictures is no longer dependent on surgical techniques alone. Percutaneous and endoscopic dilation techniques now offer additional methods of treatment and are commonly used either as an alternative to, or in conjunction with, surgical management. In this chapter the radiological assessment of benign biliary strictures is discussed and the available percutaneous radiological techniques are described in detail.

AETIOLOGY, LOCATION AND PRESENTATION

Benign biliary strictures fall into two aetiological groups: traumatic and nontraumatic.

Traumatic strictures

POSTOPERATIVE STRICTURES

The majority of benign biliary strictures in Western countries are due to upper abdominal surgery, and in particular to cholecystectomy, either open or laparoscopic. Liver transplantation also accounts for a small but important group of postoperative strictures.

BLUNT OR PENETRATING INJURY

Strictures may result directly from the injury or from its surgical management.

Nontraumatic strictures

1 Sclerosing cholangitis
2 Recurrent pyogenic cholangitis (oriental cholangiohepatitis)
3 Pancreatitis, usually chronic
4 Mirizzi syndrome. This group and the previous one are important to recognize because they may mimic malignant strictures
5 Intra-arterial chemotherapy/embolization

Strictures may involve **bile ducts** or **biliary-enteric anastomoses**. The site and number of strictures depend on the cause. The great majority of postoperative strictures are extrahepatic, involving the common duct or biliary-enteric anastomoses, although it is important to recognize that they can involve the right and left hepatic ducts or their

confluence, as well as intrahepatic and anomalous ducts (see Fig. 2.7, page 18).

Intrahepatic strictures may be secondary to trauma, operative or otherwise, but are more often due to sclerosing cholangitis, or associated with long-standing intrahepatic calculi or recurrent pyogenic cholangitis. The majority of patients present with jaundice and cholangitis, which is often recurrent, and if neglected will tend to progress to hepatic fibrosis, secondary biliary cirrhosis and portal hypertension. Strictures involving one hepatic lobe or segment tend to present with cholangitis and cholestatic liver function tests, with elevation in alkaline phosphatase and γ-glutamyl transferase, but with little or no elevation in serum bilirubin levels. Between episodes of cholangitis liver function tests may be normal or minimally abnormal.

RADIOLOGICAL ASSESSMENT

Direct cholangiography

Direct cholangiography is the most important radiological investigation and it is essential that all of the biliary tract is well demonstrated so that multiple strictures and/or associated ductal calculi are recognized. Endoscopic retrograde cholangiography (ERC) may be sufficient but may fail to opacify the proximal biliary tract adequately especially if the stricture is tight, or if there is lobar or segmental intrahepatic stricturing. Furthermore endoscopic access may be precluded by previous surgery. Percutaneous transhepatic cholangiography (PTC) is therefore frequently necessary for complete cholangiography. In cases of suspected segmental ductal stricturing, PTC may need to be ultrasound-guided to allow puncture of selected segments. Ultrasound guidance is also useful if there is marked lobar atrophy associated with ductal stricturing (Fig. 2.4, page 16).

The aetiology of stricturing is usually evident from clinical and cholangiographic information. Postoperative strictures are almost always very short, of the order of 1–3 mm in length. Benign inflammatory strictures due to sclerosing cholangitis, pancreatitis or the Mirizzi syndrome tend to be longer and may be difficult to differentiate from malignant strictures. Cytological or histological examination may be indicated if other clinical or radiological differentiating features are not helpful, and samples can be obtained by percutaneous needle biopsy (guided by cholangiography or ultrasound), or by brushings or endoluminal biopsy if there is a larger percutaneous track already established.

Ultrasound

Ultrasound is often the first radiological investigation and careful examination should evaluate the following:

1 Ductal dilation. The degree of dilation tends to be less than with malignant obstruction and may be minimal. The absence of dilation therefore does not exclude benign stricturing and this applies to postoperative strictures as well as sclerosing cholangitis. Dilation may be lobar or segmental so all of the liver must be carefully examined.
2 Evidence of portal hypertension, the presence of which may influence the choice of management.
3 The presence of hepatic abscesses which may complicate cholangitis.
4 The presence of intraductal calculi.
5 The presence of tumour as the cause of stricturing.

Computed tomography

Computed tomography (CT) should be performed if ultrasound is technically unsatisfactory or difficult to interpret, and the same factors are evaluated as listed above for ultrasound.

CONTRAINDICATIONS TO INTERVENTION

The main contraindication to transhepatic intervention is the presence of uncorrectable coagulopathy or substantial ascites. These are only relative contraindications if access for intervention is available via a superficially fixed Roux loop, or via a mature tube track.

Segmental strictures with no significant cholangitis may be managed expectantly if symptoms are minimal (Hadjis *et al.*, 1986).

PATIENT PREPARATION

1 Radiological assessment as above.
2 Clotting profile and correction of any reversable coagulopathy.
3 Intravenous line with appropriate attention to hydration and electrolyte balance.
4 Antibiotics. Broad-spectrum antibiotic cover is used routinely, for example, piperacillin, 2 g, intravenously 1 hour prior to the procedure. Antibiotic selection should cover the common Gram-positive and Gram-negative intestinal organisms which are most often responsible for cholangitis. Anaerobic cover (e.g. metronidazole) should be added in frankly septic patients or if anaerobes have been isolated from blood or bile cultures. In patients who are septic secondary to cholangitis, biliary drainage should be used for a few days with antibiotic therapy before further biliary manipulations are performed. Antibiotics should be continued after the procedure if there is clinical sepsis but otherwise this is not necessary as a routine.

ANAESTHESIA

The comments in Chapter 2 are applicable. For procedures via an existing tract or a superficially fixed Roux loop intravenous sedation and analgesia are usually adequate, although dilating strictures is painful and some patients will elect to have general anaesthesia. Patients should be carefully monitored with attention to circulatory and respiratory status.

IMAGING GUIDANCE

Procedures are performed on a fluoroscopic table with undercouch tube. Biplane or C-arm fluoroscopy is an advantage but not essential. The screening quality should be good enough to visualize small calculi. Any previous CT scans or cholangiography should be available to help determine the most appropriate percutaneous approach. Direct ultrasound guidance for initial bile duct puncture is useful in those patients with segmental obstruction, marked lobar atrophy (Fig. 2.4, page 16), or if there has been previous hepatic resection.

RADIOLOGICAL BILIARY ACCESS

The options for gaining percutaneous radiological access to the biliary tract are outlined below.

Transhepatic access (Figs 4.1 and 4.2)

This is frequently the only radiological option available. The right lobe can be approached from the right flank, and the left lobe from the epigastrium, or very occasionally from the left flank if there is marked left lobe hypertrophy. The transhepatic approach usually provides a very favourable anatomical approach to the stricture. Its disadvantages are the discomfort and morbidity of transhepatic catheterization, the inability to access multiple hepatic segments easily from a single tract, and the need to leave an indwelling catheter in place if long-term access is required.

The technique of gaining transhepatic biliary access is as described in Chapter 2.

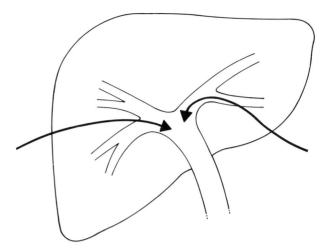

Fig. 4.1
Transhepatic biliary access. The right lobe is approached via the right flank and the left lobe via the epigastrium. If there are segmental strictures or calculi the appropriate segment is chosen for puncture if it is accessible. For left lobe puncture it is usually best to gain access via the segment 3 duct which lies anterioinferior in the lateral segments of the left lobe (see Chapter 2).

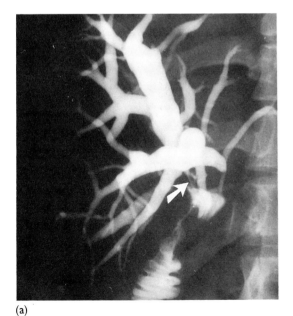

(a)

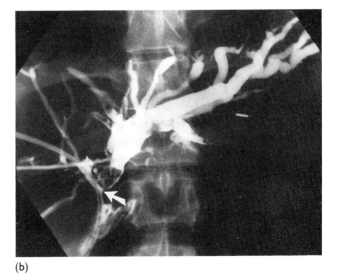

(b)

Fig. 4.2
Strictures (arrows) of (a) right and (b) left biliary-enteric anastomoses following repair of a postcholecystectomy duct injury. A calculus lies in the left hepatic duct immediately above the stricture. Percutaneous transhepatic access via the (c) right flank and (d) epigastrium was used to dilate the strictures with angioplasty balloon catheters. The left hepatic duct stone was crushed by the balloon and fragments flushed into the Roux loop. (e) The cholangiogram at 6 weeks postdilation shows no residual stricturing or stones and the percutaneous catheters were removed.

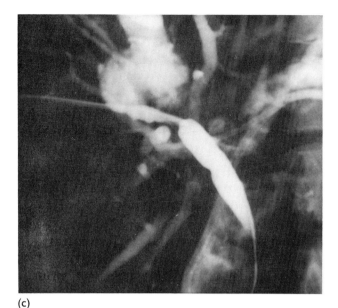

(c)

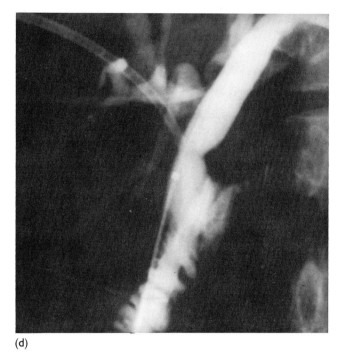

(d)

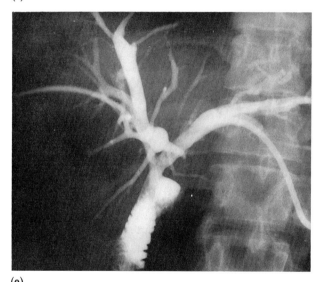

(e)

Fig. 4.2 *Continued*

Even if not used for stenting reasons, such tubes can be inserted at operation with the primary purpose of providing subsequent radiological access. This is one way of usefully combining surgical and radiological techniques in complex benign biliary problems (Gibson *et al.*, 1987). T-tubes inserted into the common bile duct can also be useful for access.

Radiological access is simply achieved by inserting a straight guidewire through the surgical tube which is then removed. A replacement tube can be inserted after the procedure to allow subsequent short- or long-term access.

Access via an operatively placed tube (Fig. 4.3)

If surgical stricture repair is used, the operative placement of a tube within the biliary tract can provide an extremely useful and atraumatic approach for subsequent radiological management of any associated intrahepatic strictures or calculi which are surgically inaccessible. These tubes are used routinely by some surgeons to stent biliary-enteric anastomoses and are sometimes brought out through the liver in the form of a U-tube (Fig. 4.3).

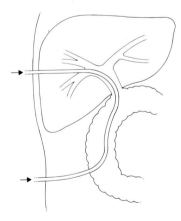

Fig. 4.3
Biliary access via an operatively placed tube. The tube will have usually been placed across the site of biliary-enteric anastomosis at the time of stricture repair and may be brought out through the liver in the form of a 'U-tube' as shown, or may have one end lying within an intrahepatic duct.

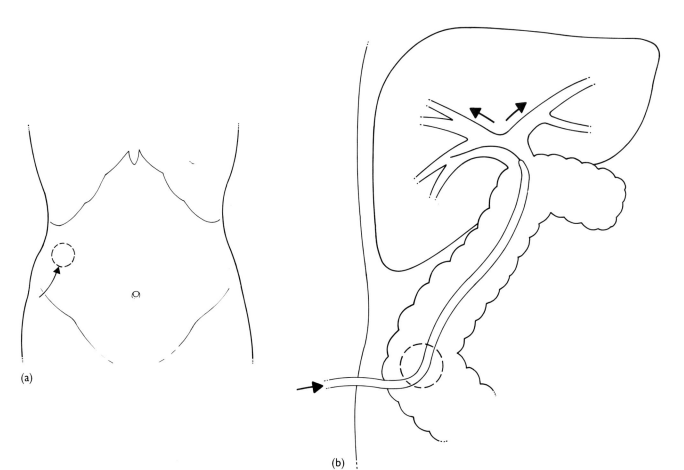

Fig. 4.4
Biliary access via a superficially fixed Roux loop ('Hutson loop'). One limb of the Roux loop used for biliary-enteric anastomosis is fixed to the anterior abdominal wall. (a) The preferred technique is for the efferent limb to be fixed securely to the parietal peritoneum inferior to the right lobe of liver towards the right iliac fossa. (b) The site of fixation is marked in some way, such as with a ring of surgical clips, allowing subsequent percutaneous puncture of the loop under fluoroscopic guidance, and retrograde catheterization of the biliary tract via the efferent limb of the Roux loop. Access can then be gained to either the right or left lobes and their segmental branches from a single percutaneous puncture.

Access via a superficially fixed Roux loop (Figs 4.4 and 4.5)

Another combined surgical–radiological approach is to provide percutaneous biliary access via a superficially fixed Roux loop (Hutson *et al.*, 1984). This approach is available where a Roux loop has been used for a biliary-enteric anastomosis and either the afferent or efferent limb of the loop is fixed to the anterior abdominal wall. The fixation must be secure and its site well marked with metal clips (Fig. 4.5a).

The best radiological approach is usually gained by fixing the efferent limb inferiorly on the right side towards the right iliac fossa (Fig. 4.4). This is preferable to fixing the afferent limb in the central abdomen as it allows comfortable radiological manipulation from the right side with the operator's hands removed from the primary X-ray beam, and

usually allows a more direct approach to both the right and left hepatic ducts.

The Roux loop approach has the major advantages of:

1 Avoiding the discomfort and morbidity associated with transhepatic catheterization.
2 Allowing access to all hepatic segments from a single percutaneous approach (Fig. 4.4b).
3 Providing long-term atraumatic radiological access for follow-up cholangiography and repeated interventions as required, without the need for indwelling catheters.

We have found this approach to be so valuable that it is now our practice to create superficial Roux loop access in almost all patients undergoing biliary-enteric anastomosis for treatment of biliary strictures or intrahepatic calculi. Furthermore, in patients who

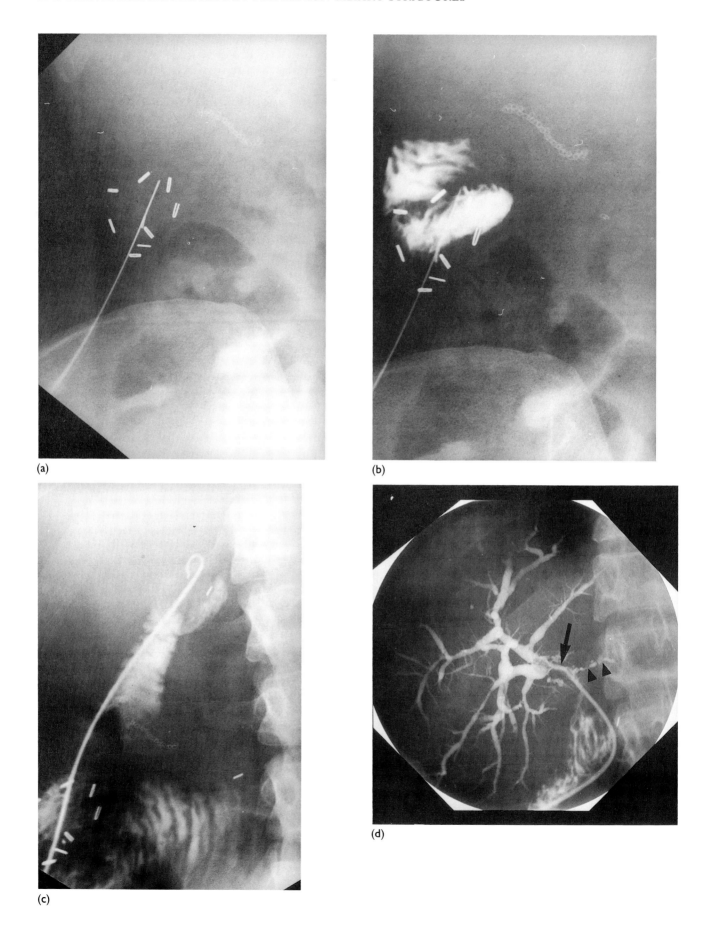

(a)

(b)

(c)

(d)

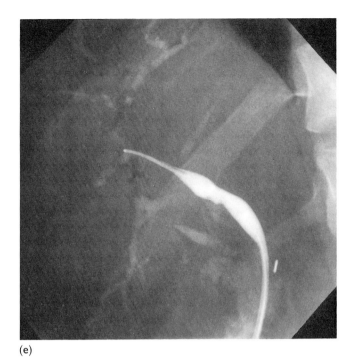

(e)

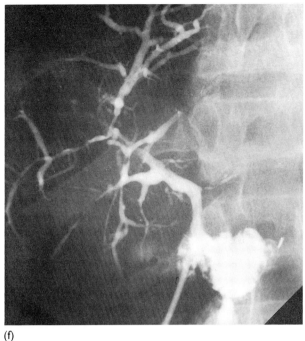

(f)

Fig. 4.5 *Biliary access via a superficially fixed Roux loop in a patient with primary sclerosing cholangitis*
(a) The Roux loop is punctured with a 22 gauge needle inserted through the centre of the ring of surgical clips marking the site of fixation. (b) Position of the needle tip is established by injection of a *small* amount of contrast medium, and the position adjusted slightly until the tip is intraluminal. Air or saline can then be injected to help maintain Roux loop distension. (c) A system such as the Neff (Cook) is used to introduce a 'J' guidewire over which a manipulation catheter (see Fig. 2.14, page 24) is passed. The combination of wire and catheter is used to negotiate the Roux loop to the site of the biliary-enteric anastomosis. Further distension and opacification of the Roux loop with dilute contrast medium via the catheter is often helpful. (d) Cholangiogram via the catheter shows a tight stricture of the right hepatic duct (arrow) with a calculus behind it. The left hepatic duct is faintly opacified (arrowheads) and the left lobe was markedly atrophic. (e) The stricture was dilated with a balloon catheter and the calculus removed with a stone basket. (f) Completion cholangiogram showing stricture dilation and stone clearance. (g) At a subsequent procedure in the same patient a 14 Fr sheath was inserted over the guidewire, to lie just below the biliary-enteric anastomosis to facilitate catheter manipulations. Reproduced from Gibson (1990) with permission of Mosby-Year Book.

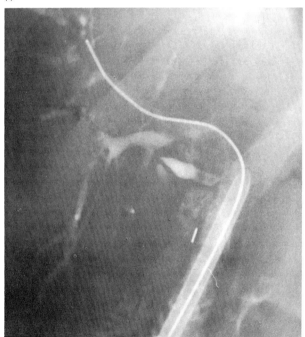

(g)

have an existing Roux loop which previously has not been fixed superficially, it may be worthwhile operating with the specific aim of fixing the Roux loop to provide radiological access for recurrent biliary strictures or intrahepatic calculi.

The Roux loop is punctured with a 22 gauge Chiba needle under fluoroscopic guidance using the surgical clips to choose the puncture site (Fig. 4.5a). Small volumes of contrast medium are injected until the needle tip is manipulated into the Roux loop

lumen (Fig. 4.5b), and an introducing system such as the Neff introduction set (Cook) is used to insert an 0.038 inch J guidewire. A biliary manipulation catheter with good torque control and a very short distal bend (Fig. 2.14 in Chapter 2) is then used with the guidewire to negotiate the bowel lumen to reach the biliary-enteric anastomosis. It is sometimes helpful to inject air or normal saline to distend the Roux loop during introduction and passage of the wire. Once the anastomosis is reached it is preferable to replace the J wire with a soft-tipped straight wire (e.g. Newton guidewire, Cook) to cross the anastomosis, controlling wire direction with the catheter, and this combination also is used to negotiate ductal strictures. As with malignant strictures a hydrophilic polymer-coated wire (Radifocus wire, Terumo) may be helpful in crossing tight strictures or negotiating tortuous duct segments. Very occasionally puncture of the Roux loop is difficult and it is necessary to perform a PTC to opacify the loop for better localization.

Fibrosis around the site of Roux loop fixation can make access difficult, and changing the site of percutaneous puncture slightly may help. Incremental dilation over a stiff 0.038 inch guidewire (e.g. Amplatz extra-stiff, Cook) is performed to allow placement of a 10 or 12 Fr sheath which facilitates subsequent catheter manipulation. Dilation of the percutaneous tract should be limited to the minimum size necessary for subsequent manipulations because of the potential risk of dislodging the Roux loop. Roux loop detachment is very uncommon if the loop has been fixed securely, and it is recognizable by intraperitonal spill of contrast medium. If this occurs a percutaneous drainage catheter with side holes should be inserted over the same guidewire into the adjacent peritoneal cavity followed by surgical repair and refixation of the loop.

Having gained access to the biliary tract it is helpful to place the upper end of the sheath close to the biliary-enteric anastomosis (Fig. 4.5g). This allows easier catheter manipulation and exchange, and is especially useful during stone extraction. For stone extractions it is occasionally necessary to dilate to a larger sheath (e.g. 16 Fr). A second guidewire can be inserted through the sheath to maintain access to a particular duct.

In creating the Roux loop the surgeon should make the access limb (between the fixation site and the biliary anastomosis) as straight and short as possible, to avoid having to attempt radiological manipulation through a long and tortuous Roux loop. The insertion of a sheath through a tortuous Roux loop can often salvage this situation but surgical revision of fixation of the loop may be necessary.

TECHNIQUES FOR STRICTURE DILATION

The majority of strictures can be dilated using angioplasty balloon catheters (Fig. 4.2) although resistant strictures may need coaxial dilation. The balloon dimensions are determined by the stricture length and the diameter of the adjacent normal duct. A 2 or 4 cm balloon length is usually adequate. Balloon diameter should approximate the diameter of adjacent normal duct with the aim of obliterating the waist of the balloon caused by the stricture. Commonly used sizes are 6 mm diameter for intrahepatic strictures and 8 to 10 mm for extrahepatic or anastomotic strictures. Inflation pressures of 10–12 atm are usual (e.g. Accent DG balloon catheter, Cook) and occasionally pressures of up to 17–18 atm are necessary (e.g. Blue Max, Medi-Tech). A mechanical inflation device (e.g. Encore inflation device, Scimed) is used for the dilation. The number and duration of inflations is empirical, the aim being to restore the lumen to the diameter of the adjacent normal duct, with no residual waisting of the balloon. Complete dilation sometimes cannot be achieved. If minimal dilation is produced by ballooning, coaxial dilation may be successful.

After dilation and/or stone removal, a temporary catheter should be left in place to allow check cholangiography at 24 to 48 hours if there is any doubt as to the adequacy of dilation or of stone clearance.

POSTDILATION STENTING

The value of postdilation stenting of strictures is uncertain, and we do not use long-term stenting routinely. A distinction should be made between placing a long-term large diameter tube (e.g. 20 Fr) across a stricture as a stent in an attempt to reduce the restenosis rate, and leaving a smaller catheter in place for a shorter period to maintain radiological access. Indwelling catheters have the advantage of allowing check cholangiography and repeat dilations if necessary. The disadvantages are that they produce discomfort and provide a persisting route for bacterial colonization, and intraductal calculus formation is probably encouraged by the presence of catheters (Schuller et al., 1981). Although long-term stenting catheters are used routinely by some, there is no convincing evidence that their use significantly reduces the restenosis rate.

Our practice is to reserve the use of long-term stenting tubes for patients who do not have superficial Roux loop access, and who have very tight strictures which are difficult to dilate, or which recur early. Otherwise, in patients without Roux loop access, we leave an 8 to 10 Fr catheter in place across the stricture postdilation and remove it if there is no evidence of restenosis on follow-up at 6–8 weeks (see below). In patients with Roux loop access we do not leave any catheter in place beyond 24 to 28 hours, and in most cases leave no catheter unless there is doubt as to the adequacy of stricture dilation or stone clearance.

METAL STENTS

Approximately 20–30% of patients with benign biliary strictures will present with restenosis following dilation (Gibson *et al.*, 1988). Patients with multiple recurrences are best dealt with by repeated dilations via a Roux loop. However, in a small number of cases the frequency of procedures required is such that the patient's quality of life becomes unacceptably low. In this small group of patients metallic endoprostheses appear to be beneficial.

Many patients with multiple stricture recurrences have had a large number of attempted surgical repairs and, as a consequence, the strictures are very difficult to dilate using balloon catheters. It is, therefore, probably inadvisable to use balloon expandable metallic stents in this situation. In addition, such stents do not exert any outward pressure on the bile duct and are less likely to maintain their position across a short stricture. In our experience, the metallic endoprosthesis best suited for the treatment of recurrent benign biliary strictures is the Gianturco stent (see Chapter 3). This exerts a constant outward radial force and can dilate strictures which cannot be dilated even with the strongest angioplasty balloons available (Fig. 4.6). The stent can be inserted transhepatically but, depending on the precise diameter of the endoprosthesis to be used (usually 8–10 mm), a 10 or 12 Fr introducer sheath is required. This usually necessitates a two-step procedure to allow a transhepatic track to form. We prefer to insert these stents retrogradely via a Roux loop if this is available. The stent, within the introducer sheath, is positioned across the stricture. Following this the sheath is withdrawn whilst maintaining the position of the endoprosthesis with a plastic pusher. Immediately after its release the stent is only partially expanded but, within a few weeks, it expands to achieve its maximum diameter (Fig. 4.6).

Results of three early trials using Gianturco stents in benign biliary strictures are summarized in Table 4.1 (Coons, 1989; Irving *et al.*, 1989; Rossi *et al.*, 1990). Although several stents have been misplaced

Table 4.1 *Gianturco stents in benign biliary strictures.*

	Patients	Follow-up	Dislocation %	Obstruction %
Coons, 1989	15	6	0	13
Rossi, 1990	17	8	12	12
Irving, 1989	11	6–21	7	7
Total	43	7	6	11

at the time of stent insertion, only one delayed stent migration has been reported (Irving *et al.*, 1989). This was a single nonbarbed Gianturco stent placed across a short anastomotic stricture, which migrated into the jejunal loop after 5 months. We have recently observed two late stent migrations in a patient with a similar stricture. Both the stents were barbed unmodified Gianturco stents. These were replaced with Rosch/Gianturco stents which have maintained their position very well. This version of the Gianturco stent is probably more suited to short strictures than was the initial design as the waist between two stents can be placed precisely across the stricture thus preventing any movement of the endoprosthesis (Fig. 4.7).

Gianturco stent occlusion has been reported in 7–13% of cases. This is thought to be due to mucosal hyperplasia or fibrotic overgrowth in most instances, although a few cases have been due to stricture recurrence between the junction of two single stents. No serious complications have been seen and the 30-day mortality has been zero. This compares favourably with a mortality rate of up to 7.7% attributed to surgery (Pitt *et al.*, 1981). The long-term patency rate and the effect of metallic stents on the bile duct have not been documented. Experience with the Wallstent endoprosthesis in benign strictures is more limited than with the Gianturco stent. The Wallstent maintains its position very well and appears promising in the management of complex strictures. Although the initial results appear promising, until the questions concerning long-term outcome are resolved, metallic biliary stents should only be used in the treatment of benign biliary strictures in selected cases where repeated balloon dilation has failed.

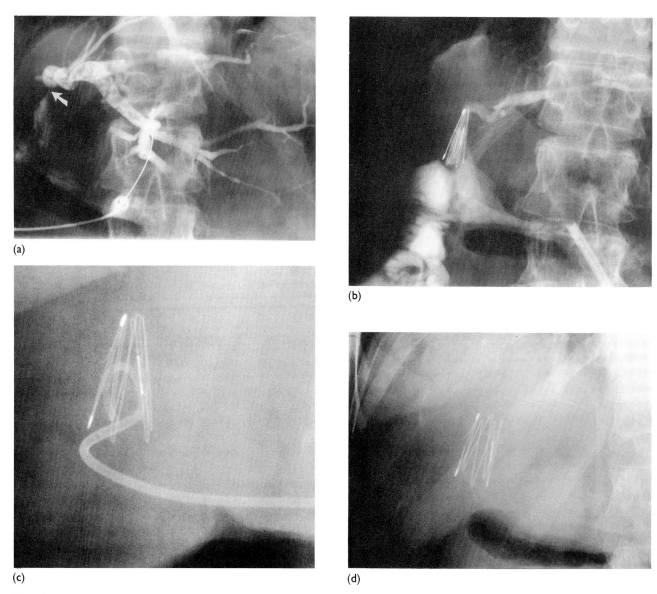

Fig. 4.6
(a) Severe stricture (arrow) at the site of the anastomosis between the left hepatic duct and the loop of jejunum. The intrahepatic ducts contain numerous calculi. (b) Following moderate dilation of the stricture and clearing of the calculi using saline irrigation a Gianturco stent was inserted across it. This is still only partially expanded. (c) The stent expanded a little more 7 days later. (d) The stent expanded completely 1 month after the insertion.

FOLLOW-UP

Short-term follow-up

If an indwelling catheter is in place direct cholangiography is the simplest method of checking on progress and this should be performed at 6 to 8 weeks postdilation. If there is any apparent stricturing the degree of stenosis is better assessed by gently inflating a balloon across the stricture and observing any waisting of the balloon. If there is a significant residual stenosis then repeat dilation is attempted and a catheter left in place for a further few weeks. When the stricture size is unchanged on two consecutive balloon measurements, or if there is no residual stenosis then the stricture is regarded as stable and the catheter removed. In difficult cases where there is doubt about the adequacy of dilation it may be helpful to 'challenge' the stricture prior to catheter removal (Williams *et al.*, 1987), but we do not use this approach routinely. The 'challenge' is performed by retracting the catheter so that it lies above the stricture and observing the patient for a few weeks for clinical or biochemical evidence of residual or recurrent stricturing.

(a)

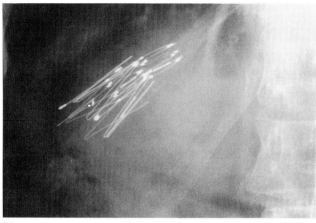

(b)

Fig. 4.7
(a) Following the migration of two unmodified Gianturco stents below a strictured hepaticojejunostomy a Rosch/Gianturco stent has been inserted across the stricture. This is still only partially expanded. (b) A radiograph taken a few months later shows expansion of the stent.

Long-term follow-up

Long-term follow-up is essential as there is a significant restenosis rate following stricture dilation. Experience with restenosis following surgical repair indicates that approximately 20% of recurrent strictures do not manifest until after 5 years (Pitt *et al.*, 1982). Follow-up should include clinical, biochemical and radiological reviews.

CLINICAL REVIEW

Features suggesting inadequate dilation or stricture recurrence are jaundice or cholangitis. The latter classically presents as pain, fever and jaundice, but frequently there will be only intermittent pain or fever associated with abnormal liver function tests without clinical jaundice, especially if obstruction is lobar or segmental.

BIOCHEMICAL REVIEW

Providing there is no underlying diffuse hepatic disease, serum bilirubin should fall to normal levels. Alkaline phosphatase levels should also fall, although occasionally they will remain slightly elevated despite other evidence of adequate dilation (Benjamin, 1988). If, however, the degree of elevation is marked or is rising, then stricture recurrence must be suspected. Lobar or segmental ductal obstruction can result in high alkaline phosphatase levels with normal serum bilirubin.

RADIOLOGICAL REVIEW

If there is reason to suspect stricture recurrence on the above criteria then direct cholangiography is indicated. Direct cholangiography can also be performed periodically if long-term stenting catheters are used. If access is available via a superficial Roux loop then cholangiography can be safely and simply performed by percutaneous puncture of the loop and retrograde catheterization of the biliary tract.

We no longer use ultrasound, CT or HIDA scanning for routine follow-up as each of these has its pitfalls in detection of stricture recurrence; we rely instead on clinical and biochemical review, and direct cholangiography as indicated. Ultrasound or CT can be helpful, however, in evaluating recurrent segmental or lobar obstruction when suspected on clinical or biochemical grounds.

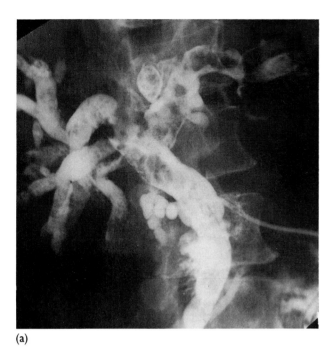

(a)

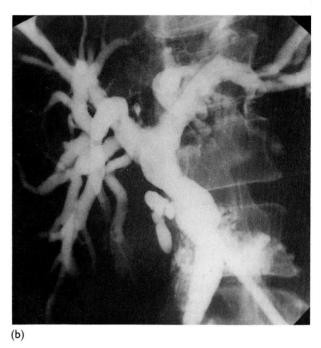

(b)

Fig. 4.8
Extensive intrahepatic and extrahepatic ductal calculi associated with Oriental cholangiohepatitis. Choledochojejunostomy was performed with superficial Roux loop fixation (in this case the fixation was in the central abdomen). (a) Postoperatively numerous calculi remained. (b) Using a combination of basket extraction, crushing and flushing, stone clearance was achieved using percutaneous Roux loop access.

MANAGEMENT OF ASSOCIATED CALCULI

(see also Chapter 1)

Intraductal calculi are often associated with benign strictures and are most often intrahepatic; it is essential to make a thorough search for intrahepatic calculi at the time of stricture dilation. All segments should be opacified and care should be taken not to overlook a calculus obstructing the orifice of a segmental duct. Prior to calculus extraction any strictures should be dilated to aid access and stone clearance.

The options for removing stones depend on stone location, size and composition, as well as the radiological access. The following discussion regarding available options also applies to patients in whom the major problem is intrahepatic stones rather than strictures, such as in Oriental cholangiohepatitis (Fig. 4.8), although these patients may also have strictures (Chetty *et al.*, 1990).

Basket extraction

Calculi can be removed intact or after fragmentation and because ductal calculi tend to be pigment stones they are often very easily fragmented. Fragments can then be basketed or flushed from the ducts. Tougher large stones require a sturdier stone-crushing basket, which require a larger tract or Roux loop approach. Basket design and use is discussed in Chapter 1.

Perfusion

The results for dissolution of ductal calculi are disappointing and only partial success can be achieved despite long perfusion periods (Butch *et al.*, 1984). This is not surprising as dissolution agents currently available, for example, mono-octanoin acid, rely on a high cholesterol content in calculi whereas ductal stones associated with strictures are mainly pigment stones.

It is, however, sometimes helpful to perfuse the ducts slowly with normal saline to clear multiple small stones or fragments which otherwise cannot be removed. A catheter with distal side holes is placed above the stones and secured well at the skin. A 12 to 24 hour perfusion of 1 or 2 litres of saline is then commenced, and maintained until check cholangiography the following day; the patient should be closely observed and if pain develops the perfusion is ceased.

Lithotripsy

Extracorporeal shockwave lithotripsy (ESWL) using fluoroscopic guidance can be valuable for fragmenting larger stones in the extrahepatic bile ducts, with subsequent passage or extraction of the fragments (Becker *et al.*, 1987).

If a sufficiently large tract is created then choledochoscopy can be used in conjunction with ultrasound or electrohydraulic lithotripsy.

COMPLICATIONS

Complications of stricture dilation depend on the access used. Those associated with a transhepatic approach are similar to those seen with transhepatic drainage for malignant biliary obstruction, namely, haemorrhage, bile leak and septic complications (Chapter 2). With any of the approaches cholangitis and septicaemia are not infrequent complications. Routine antibiotic cover (see above) and postprocedural observation are essential.

Dislodgement of superficially fixed Roux loops can occur if surgical fixation is not secure or if dilation of the percutaneous tract is over vigorous; dilation to 12–14 Fr is usually safe and adequate.

Late complications include stricture and/or stone recurrence, and occlusion or dislodgement of long-term catheters.

RESULTS

In the series reported up to 1988, stricture dilation was deemed successful in 72% of 249 patients at a mean follow-up of 30 months (Gibson *et al.*, 1988). Stricture recurrence occurred in up to 34% of patients in these series. There was no significant difference between anastomotic and nonanastomotic strictures, with the exception of patients with primary sclerosing cholangitis (PSC) of whom only 59% of patients had a good result. Patients with PSC, however, often present major problems for surgical management and dilation of dominant strictures can be valuable particularly if long-term access is available via a Roux loop (Fig. 4.5).

Longer term follow-up results are scarce and it is almost certain that longer follow-up will reveal higher rates of stricture recurrence. Pitt *et al.* (1989) reported a 55% success rate for stricture dilation in a group of 20 patients with a mean follow-up of 5 years. Nevertheless, the results available for stricture dilation are very respectable especially given that most of the series reported include a substantial proportion of patients with complex biliary problems unsuitable for a surgical approach alone.

ROLE OF RADIOLOGICAL INTERVENTION IN RELATION TO SURGICAL MANAGEMENT

In the absence of controlled trials (which almost certainly will never be performed), and given the variability in local availability of surgical and radiological expertise, it is difficult to be dogmatic about the relative place of interventional radiological and surgical management for benign biliary strictures. The best results for surgical treatment are better than any published results for radiological management and in specialist surgical centres the treatment of choice for strictures of the extrahepatic ducts should be surgery. Apart from local expertise the following factors adversely affect outcome of surgical stricture repair (Blumgart *et al.*, 1984):

1 Multiple previous operative attempts at repair
2 Stricture involvement of the right and left hepatic ducts or their confluence
3 Intrahepatic strictures and calculi
4 Presence of right lobe atrophy and left lobe hypertrophy (the atrophy/hypertrophy complex)
5 Portal hypertension

The presence of one or more of these factors should favour selection of a radiological approach either alone or in combination with surgery.

REFERENCES

BECKER CD, FACHE JS, GIBNEY RG, SCUDAMORE CH, BURHENNE HJ (1987). Choledocholithiasis: treatment with extracorporeal shock wave lithotripsy. *Radiology* **165**: 407–8.
BENJAMIN IS (1988). Biliary tract obstruction. In: *Surgery of the Liver and Biliary tract*, pp. 111–19. Edited by Blumgart LH. Churchill Livingstone, Edinburgh.

BLUMGART LH, KELLY CJ, BENJAMIN IS (1984). Benign bile duct stricture following cholecystectomy: critical factors in management. *British Journal of Surgery* **71**: 836–43.

BUTCH RJ, MACCARTY RL, MUELLER PR, FERRUCCI JT, SIMEONE JF, TEPLICK SK, HASKIN PH (1984). Monooctanoin perfusion treatment of intrahepatic calculi. *Radiology* **153**: 375–7.

CHETTY MN, YEUNG EYC, BENJAMIN IS, ADAM A (1990). A strategy for the percutaneous removal of multiple intrahepatic biliary calculi. *Journal of International Radiology* **5**: 167–70.

COONS HG (1989). Self-expanding stainless steel biliary stents. *Radiology* **170**: 979–83.

GIBSON RN (1990). Interventional radiology for benign biliary strictures. In: *Advances in Hepatobiliary Radiology*. Edited by Ferrucci JT and Mathieu DG. Mosby Year Books, St louis.

GIBSON RN, ADAM A, CZERNIAK A, HALEVY A, HADJIS N, BENJAMIN IS, ALLISON DJ, BLUMGART LH (1987). Benign biliary strictures: a proposed combined surgical and radiological management. *Australian and New Zealand Journal of Surgery* **57**: 361–8.

GIBSON RN, ADAM A, YEUNG E, SAVAGE A, COLLIER NA, BENJAMIN IS, BLUMGART LH, ALLISON DJ (1988). Percutaneous techniques in benign hilar and intrahepatic benign biliary strictures. *Journal of Interventional Radiology* **3**: 125–30.

HADJIS NS, CARR D, BLENKHARN I, GIBSON R, BLUMGART LH (1986). Expectant management of patients with unilateral hepatic duct stricture and liver atrophy. *Gut* **27**: 1223–7.

HUTSON DG, RUSSELL E, SCHIFF E, LEVI JJ, JEFFERS L, ZEPPA R (1984). Balloon dilation of biliary strictures through a choledochojejuno-cutaneous fistula. *Annals of Surgery* **199**: 637–44.

IRVING JD, ADAM A, DICK R, DONDELINGER RF, LUNDERQUIST A, ROCHE A (1989). Gianturco expandable metallic biliary stents: results of a European clinical trial. *Radiology* **172**: 321–6.

PITT HA, CAMERON JL, POSTIER RG, GADACZ TC (1981). Factors affecting mortality in biliary tract surgery. *American Journal of Surgery* **141**: 66–70.

PITT HA, MIYAMOTO T, PARAPATIS SK, TOMPKINS RK, LONGMIRE WP (1982). Factors influencing outcome in patients with post-operative biliary strictures. *American Journal of Surgery* **144**: 14–21.

PITT HA, KAUFMAN SL, COLEMAN J, WHITE RI, CAMERON JL (1989). Benign postoperative strictures. Operate or dilate? *Annals of Surgery* **210**: 417–25.

ROSSI P, BEZZI M, SALVATORI FM, MACCIONI F, PORCARO ML (1990). Recurrent benign biliary strictures: management with self-expandable metallic stents. *Radiology* **1975**: 661–5.

SCHULLER AM, REZK GJ, LYON DT (1981). Calculus formation around common bile duct stents: a complication of long term biliary drainage. *Gastrointestinal Endoscopy* **37**: 581–2.

WILLIAMS HJ, BENDER CE, MAY GR (1987). Benign post-operative biliary strictures: dilatation with fluoroscopic guidance. *Radiology* **163**: 629–34.

CHAPTER 5

Percutaneous gall bladder procedures

Eugene Y. Yeung

Introduction   68

Percutaneous cholecystostomy for acute
cholecystitis   68

Percutaneous gall stone management   71

References   79

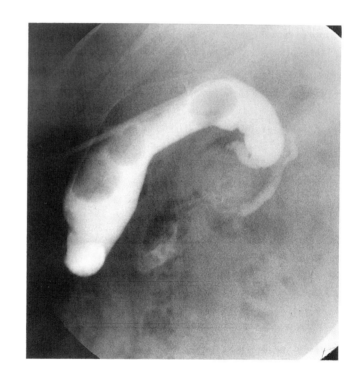

INTRODUCTION

Surgical cholecystectomy and cholecystostomy are well-established methods of treating gall bladder disease. Surgical cholecystectomy has an impressive safety record and is extremely effective. In the USA alone, more than 500 000 cholecystectomies are performed each year with a mortality of less than 1% (McSherry, 1981).

In recent years, less invasive techniques have been developed as alternatives to conventional surgery. These include laparoscopic cholecystectomy, extracorporeal shockwave lithotripsy (ESWL), percutaneous radiological cholecystostomy, cholecystolithotomy and associated techniques [e.g. methyl tert-butyl ether (MTBE) dissolution]. However, most of these alternatives can be used only if specific criteria are met, and the safety and long-term efficacy of each method still require evaluation. Laparoscopic cholecystectomy has now become the routine method of gall bladder removal in most centres. Extracorporeal shockwave lithotripsy requires a functioning gall bladder and is limited by the size and number of stones (Rawat and Burhenne, 1990). Extracorporeal shockwave lithotripsy and percutaneous radiological techniques leave the gall bladder intact, allowing new gallstones to form (Gibney et al., 1989). Research into methods of nonsurgical gall bladder ablation is, however, under way (Becker et al., 1988).

Surgery probably using laparoscopic techniques, will be the mainstay of treatment for the foreseeable future. Nevertheless, referrals for the alternative procedures are increasing. This chapter discusses some percutaneous radiological techniques currently available for treating acute and chronic gall bladder disease which offer an alternative to surgery in certain situations.

PERCUTANEOUS CHOLECYSTOSTOMY FOR ACUTE CHOLECYSTITIS

Indications and contraindications

Percutaneous radiological cholecystostomy is most often performed in patients with acute cholecystitis in whom medical therapy has failed and surgical intervention is considered to be too risky. Thus, patients who are poor anaesthetic or surgical risks and those with multisystem problems (e.g. those in intensive care) are prime candidates.

There are few contraindications to this procedure. Abnormal coagulation is a relative contraindication, although the indications for this potentially lifesaving procedure may be stronger. Another contraindication is the absence of a safe access route (e.g. gall bladder surrounded by bowel).

Patient preparation

A coagulation profile is obtained. An intravenous (i.v.) line is inserted for the administration of broadspectrum antibiotics, opiate analgesia and an antiemetic. Clotting factor replacement is given if required.

Atropine may be administered intramuscularly before or i.v. during the procedure if bradycardia is observed. It counteracts vasovagal reactions, which have been reported especially in patients with a previous history of myocardial disease (van Sonnenberg et al., 1984).

Technique

The procedure should be performed in the fluoroscopy suite, preferably using ultrasound guidance for the initial puncture. In a critically ill patient, ultrasound guidance alone may be used at the bed side.

Transhepatic and subcostal routes have been described for the gall bladder puncture. The advantages and disadvantages of each are compared in Table 5.1.

For the transhepatic approach, the needle is inserted through the right lobe of the liver horizontally and in the transverse plane or angled slightly cephalad to enter the gall bladder at the junction of its first and middle thirds. Ultrasound is used continuously to monitor needle advancement so that intrahepatic biliary and venous structures are avoided (Fig. 5.1). The transhepatic route is thought

Fig. 5.1 *Transhepatic approach*
(a) The 21/22 G needle has punctured the gall bladder under ultrasound guidance. A 0.018 inch guidewire has been inserted. The ultrasound transducer may be placed either beside the needle and between the ribs (A) or subcostally in a transverse plane (B) for imaging. (b) The Cope introduction system cannula has been inserted over the guidewire. (c) The guidewire has been withdrawn leaving the Cope system cannula in place. An 0.038 inch J-wire has been inserted and exits the cannula through its side hole. (d) The cannula has been removed, the tract dilated and a self-retaining Cope loop catheter inserted into the gall bladder.

Table 5.1 *Comparison of the Relative Advantages and Disadvantages of transhepatic and Subcostal Approaches to the Gall Bladder*

	Transhepatic Approach	**Subcostal Approach**
Advantages	Less potential for bile leakage Less risk of loss of access due to close apposition of gall bladder to liver	More direct approach with little risk to interposed structures (if ultrasound guidance is used) All parts of the gall bladder are accessible (important during cholecystolithotomy)
Disadvantages	Risk of injury to intrahepatic biliary and vascular structures, especially if the tract is enlarged for subsequent cholecystolithotomy The fundus of the gall bladder is less accessible (important during cholecystolithotomy)	More risk of intraperitoneal spillage of bile More risk of loss of access during dilator/catheter insertion

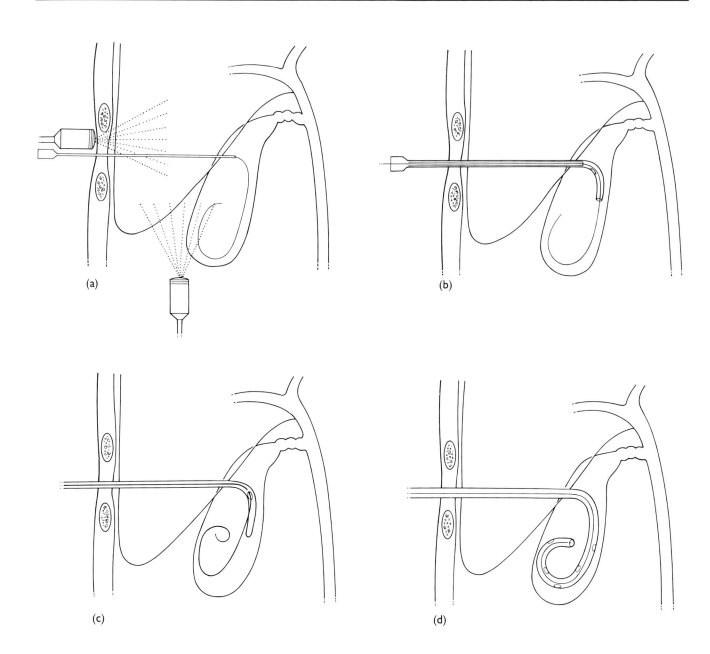

(a)

(b)

(c)

(d)

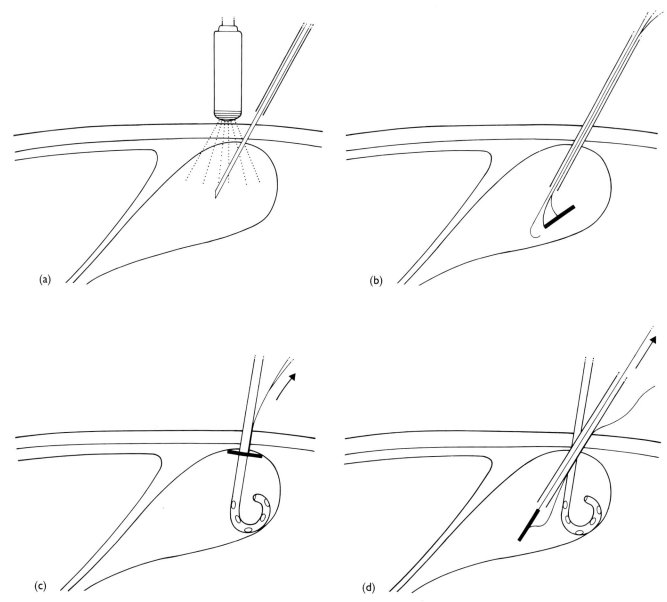

Fig. 5.2 *Subcostal approach with suture anchors (Cope, 1988)*
(a) Using ultrasound guidance, the fundus of the gall bladder is entered with a 21/22 G needle that has a removable hub (Cook). (b) A 17G needle is inserted over this. A guidewire is inserted through the needle to push out the suture anchor. (c) The suture anchor is retracted, apposing the gall bladder to the anterior abdominal wall. After serial dilation, the drainage catheter is inserted. (d) The suture anchor is removed by inserting a 16 G sheath. Pulling the thread attached to the end of the anchor allows withdrawal through the sheath.

to be safer, with less potential for bile leakage because of close apposition of the liver to the gall bladder in this region. Theoretically, loss of access from retraction of the gall bladder during catheter insertion is less likely.

In the subcostal approach, the fundus of the gall bladder is entered (Fig. 5.2). This approach is less favoured because the peritoneal cavity is traversed, risking intraperitoneal leakage of infected bile during and after the procedure, especially if the catheter is accidentally dislodged. A removable suture-anchoring device (Cope, 1988) may be used to appose the gall bladder fundus to the anterior abdominal wall during catheter insertion to give added security. However, there are many reports of successful subcostal cholecystostomy insertions without the use of suture anchors (van Sonnenberg *et al.*, 1986; Vogelzang and Nemcek, 1988; Teplick *et al.*, 1990).

With either approach, the gall bladder is best punctured with one of the commercially available introduction systems that use a 21–22 G fine needle for the initial puncture [Cope or Hawkins introduction systems; 22/18 G Mitty-Pollack (Cook, Bloomington, IN); Accustick (Meditech, Watertown, MA)].

After the needle's position in the gall bladder has been confirmed, bile is aspirated. If the aspirate is not noticeably purulent, it is sent for immediate Gram staining. If the Gram stain is negative, the gall bladder is completely decompressed with further aspiration and the needle is then removed. If infected bile is aspirated, a small amount of contrast medium is injected and the various components of the introduction system are inserted, finally leaving an 0.038 inch guidewire in place.

The tract is dilated swiftly over the guidewire and the drainage catheter is inserted. An 8–10 Fr self-retaining Cope loop-type drainage catheter with an internal stiffening cannula is recommended. Dilator/ catheter exchanges should be kept to a minimum because of the risk of bile leakage. Also, entry into the gall bladder with the dilator or drainage catheter may require a sharp thrust, especially if the gall bladder wall is thickened. Buckling of the guidewire should be particularly watched for as this may lead to loss of access.

Once the drainage catheter has been inserted, the gall bladder should be gently emptied of bile. A catheter cholecystogram to assess the gall bladder and cystic duct should not be performed at this stage.

Aftercare

Vital signs should be assessed for 24 hours after the procedure. Antibiotics should be continued. The patient's condition should improve within 24 to 48 hours. Failure to improve suggests the presence of gall bladder necrosis, in which case urgent laparotomy may be required. Another possibility is that a site other than the gall bladder may be the source of sepsis.

Once the acute septic episode has subsided, a catheter cholecystogram should be performed. If gall stones are present or the cystic duct is obstructed, the catheter should remain in place until definitive treatment in the form of surgical cholecystectomy or percutaneous cholecystolithotomy is performed.

If no gall stones are detected and the cystic duct is patent, the catheter is removed, as long as sufficient time has elapsed after insertion to allow track formation (usually 1 week). All bile should be aspirated from the gall bladder before catheter removal.

Results and complications

The technical success rate of percutaneous cholecystostomy approaches 100% (Teplick et al., 1990). The clinical success rate is difficult to determine since most patients are critically ill with multisystem failure. Percutaneous cholecystostomy should be judged clinically successful if it allows the patient's condition to improve so that definitive treatment (e.g. cholecystectomy) may be performed later with less risk.

Leakage of bile around the drainage catheter or following loss of access during the procedure is the most feared complication since it may lead to bile peritonitis. Review of the literature reveals a 1.5% risk of bile peritonitis with one death reported (Pearse et al., 1984). Some comfort may be drawn, however, from reports of catheters accidentally removed shortly after insertion without complication (Pearse et al., 1984; Larssen et al., 1988). Other potential complications include injury to venous or biliary structures by use of the transhepatic route. There is also the risk of viscus perforation with both the transhepatic and subcostal routes, although this should be low when real-time ultrasound guidance is used. Other reported complications include vasovagal reactions (van Sonnenberg et al., 1984) and haemobilia.

PERCUTANEOUS GALL STONE MANAGEMENT

Percutaneous cholecystolithotomy

The role of percutaneous cholecystolithotomy in the management of gall stones is still being defined. Although many patients would be suitable, only a relatively small proportion of patients undergo these procedures.

INDICATIONS

Patients referred for percutaneous cholecystolithotomy include those who are not suitable for surgery (e.g. high anaesthetic or surgical risk), and those who have had a surgically or percutaneously placed cholecystostomy for acute cholecystitis. All sizes, numbers and types of gall stones are suitable for this procedure.

The contraindications are similar to those for percutaneous cholecystostomy.

PATIENT PREPARATION

Preparation is as described above for cholecystostomy (see page 68).

TECHNIQUE

Percutaneous cholecystostomy

Percutaneous cholecystostomy is performed in patients who do not have an existing cholecystostomy tube. The approach to the gall bladder may be transhepatic or subcostal. Since the tract may require dilation to 24 Fr, a subcostal route may be preferred because of the potential dangers of dilating a transhepatic track to such a large size. The removal of gall stones is also technically easier through the more direct subcostal approach. On the other hand, the subcostal route has a theoretically higher risk of bile leakage. No significant difference in complications between these two approaches has been shown (Picus *et al.*, 1989; Chiverton *et al.*, 1990; Cope *et al.*, 1990).

The techniques of percutaneous cholecystostomy insertion have been described above. Usually ultrasound guidance is used, but fluoroscopy may be used if ultrasound is not available. The gall bladder is made visible fluoroscopically by ingestion of oral cholecystographic contrast medium the night before the procedure. With fluoroscopy, more needle passes will probably be needed to puncture the gall bladder and the risk of puncturing undetected interposed bowel is higher.

Track dilation

The track may be dilated to the required size in one or multiple stages over a period of days. Track size depends on the size of the gall stones but dilation to at least 18–24 Fr is usually required. If the subcostal approach is chosen and cholecystostomy, stone extraction and dilation are performed in a single session, gall bladder fixation to the anterior abdominal wall using suture anchors is recommended to prevent loss of access during dilation (Cope *et al.*, 1990) (Fig. 5.3).

In an approach with multiple stages, at least 4 weeks should elapse between cholecystostomy insertion and stone extraction to allow for track maturation. The cholecystostomy track is gradually dilated by 4 to 6 Fr every 2–3 days until the appropriate size is reached. Fixation of the gall bladder is not usually necessary (Finnegan *et al.*, 1991).

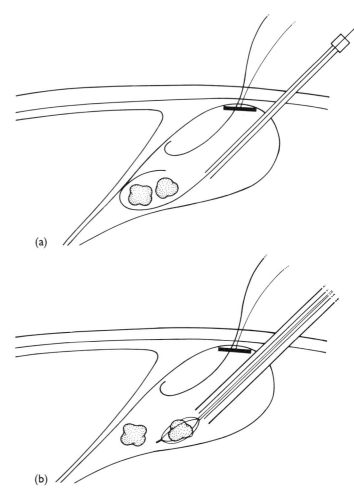

Fig. 5.3 *Subcostal cholecystolithotomy using suture anchors (Cope et al., 1990)*
(a) A suture anchor is first inserted to stabilize the gall bladder. A safety wire or catheter may be left beside the anchor. A second puncture is made 1–2 cm away. (b) The second puncture has been dilated to the appropriate size using serial dilators or a track dilating balloon. A sheath has been inserted and extraction devices are manipulated through it.

Extraction techniques

Stones are usually removed with a combination of fluoroscopic guidance and direct visualization through a flexible or rigid choledochoscope (Fig. 5.4).

The 16 Fr flexible choledochoscope (Olympus CHF-P10, Olympus Co, New Hyde Park, NY) is similar to a paediatric bronchoscope and can be inserted through an 18 Fr sheath. It has one working channel through which stone baskets or the electrohydraulic lithotriptor can be used and irrigation fluid simultaneously infused. Rigid choledochoscopes (e.g. Stortz, Kernen, Germany) available in 18 and 26 Fr sizes are essentially those used for percutaneous renal stone extraction. Rigid stone graspers, crushers and lithotripsy devices can be inserted through the

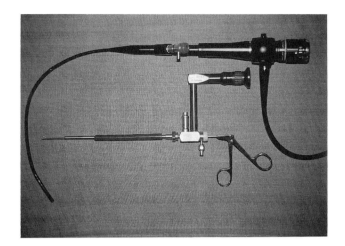

Fig. 5.4 *Choledochoscopes*
A 16 Fr flexible scope (Olympus CHF-P10, Olympus, New Hyde Park, NY) and a 26 Fr rigid scope (Stortz, Kernen, Germany) with stone forceps inserted.

working channel of these scopes. Although the use of smaller scopes seems preferable because they can be inserted through an 18 Fr track, in practice, the field of view offered by these instruments is quite limited and the larger 26 Fr rigid scope is often required. Before extraction, a safety wire should always be inserted into the gall bladder and, preferably, through the cystic duct into the common bile duct, although it may interfere with the procedure and have to be withdrawn. Stones are extracted through a sheath (usually 18–26 Fr Amplatz or peelaway sheaths). Small stones may be removed intact but large stones will require fragmentation (Figs 5.5 and 5.6).

Devices for stone extraction include:

1 Stone removal baskets passed through the endoscopes or under fluoroscopic guidance through a steerable catheter (Meditech). Baskets are available in different sizes and shapes, and with different numbers of wires, for example, four or eight prongs. If the standard baskets do not open properly because the space in the gall bladder is restricted, the shape-reforming Nitinol basket (Cook), designed specifically for such situations, may be used (Cope, 1990) (see Chapter 1, page 7, Fig. 1.4).
2 Forceps, preferably used under direct vision to guard against grasping gall bladder mucosa. If forceps are used with fluoroscopic guidance only, the technique shown in Fig. 5.7 is recommended. Various types of forceps are available, for example, grasping, crushing and Mazzariello (see Chapter 1, page 8, Fig. 1.7).
3 Irrigation, which is often effective in removing debris and dislodging adherent or impacted

stones. This is performed by inserting a curved-tipped catheter, for example, 6.5 Fr biliary manipulation catheter (Cook), through the sheath and manually injecting saline. Excess fluid should easily drain out through the sheath avoiding dangerous build up of pressure within the gall bladder. Alternatively, a pulsed jet irrigation system may be used (Cesarani *et al.*, 1988) by connecting a household dental irrigation system, for example, a WaterPik (Teledyne, Los Angeles, CA), to the catheter. This can be extremely effective in dislodging adherent stones.

Impacted stones in the cystic duct may be difficult to remove. A straight-tipped wire must be negotiated past the stones and an occlusion balloon or wire basket used (Fig. 5.8). The use of an hydrophilic-coated guidewire [Glide wire, Terumo (Meditech)] is often successful. If this fails, forceful irrigation may be attempted. Otherwise, removal may be postponed until a later session to allow oedema to decrease.

Stone fragmentation
A number of fragmentation techniques are currently available. Mechanical fragmentation may be performed under fluoroscopic guidance using stone baskets. Special stone-crushing baskets with a steel ring tip may be used (Wilson Cook, Winston-Salem, NC). Five minutes of traction on a stone trapped in these baskets is usually sufficient to fragment it. Alternatively, mechanical lithotripsy devices may be inserted along the basket to crush the stone, for example, as described by Ho *et al.* (1987) or using the Soehendra lithotriptor (Wilson Cook) (see Chapter 1, pages 8–9 Figs 1.7 and 1.8).

The other methods of lithotripsy require direct visualization of the stone for precise targeting and to prevent injury to the gall bladder wall. Electrohydraulic lithotripsy (EHL) is probably the easiest to use. A small (3 Fr), flexible copper wire electrode can be inserted through a flexible or rigid choledochoscope (Yoshimoto *et al.*, 1989). The wire electrode is placed against the stone and activated to produce a spark at its tip that generates shock waves that shatter the stone. Electrohydraulic lithotripsy generators are small, portable and relatively inexpensive, and are available in most hospitals.

Alternatively, ultrasonic lithotripsy may be used (Hwang *et al.*, 1987) although the probe is rigid and can be inserted only through a rigid scope. Laser lithotripsy uses a tunable dye laser through a small, flexible optical fibre (Bogan *et al.*, 1990). Although it is effective, the laser source is usually large and expensive.

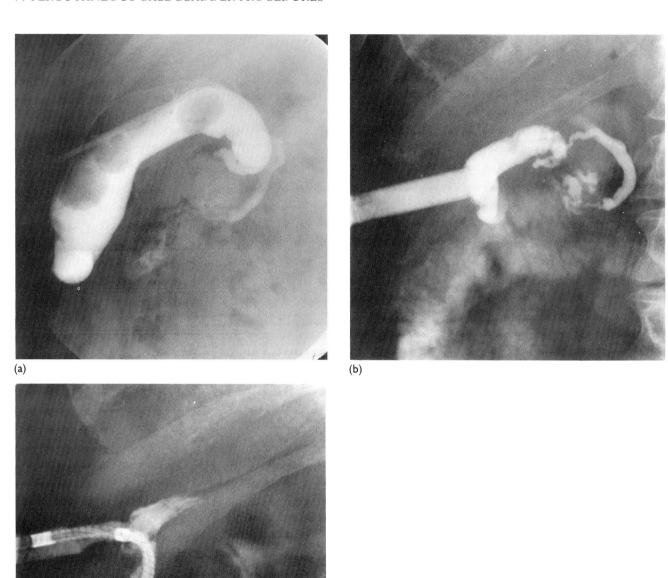

(a)

(b)

(c)

Fig. 5.5 *Transhepatic cholecystolithotomy*
(a) Transhepatic percutaneous cholecystostomy. (b) The tract is dilated to 26 Fr and stones are removed using fluoroscopy or direct vision. (c) A 16 Fr flexible choledochoscope has been inserted to inspect the gall bladder after stone clearance. Note the acute angle necessary to view the fundus in this approach. The safety guidewire was removed.

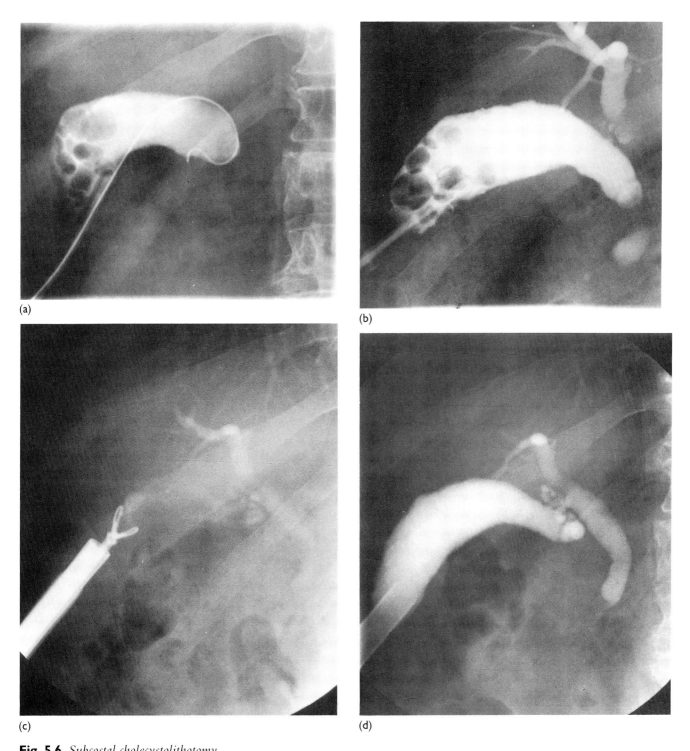

(a)

(b)

(c)

(d)

Fig. 5.6 *Subcostal cholecystolithotomy*
(a) Percutaneous cholecystostomy with the guidewire coiled in the gall bladder. (b) The drainage catheter is left in the gall bladder and the tract dilated for 4 weeks to allow tract maturation. (c) Stones are removed through a 26 Fr sheath with forceps using fluoroscopy or choledochoscopes. The safety guidewire was removed. (Note the orientation of the forceps' jaws.) (d) Clearance of stones. Note the patency of the cystic duct, which allows the sheath to be removed.

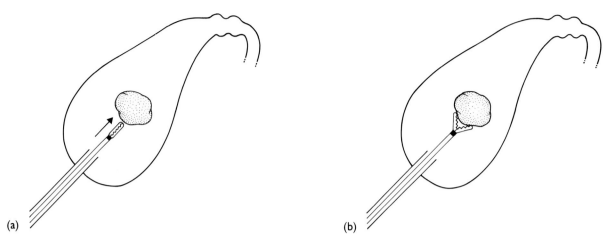

Fig. 5.7 *Use of forceps with fluoroscopic guidance*
(a) The forceps is advanced while closed until it impinges on the stone, as observed fluoroscopically. (b) The jaws of the forceps are oriented so that they open perpendicular to the X-ray beam. This reduces the risk of inadvertently grasping wall mucosa. Large stones may be fragmented before removal through the sheath.

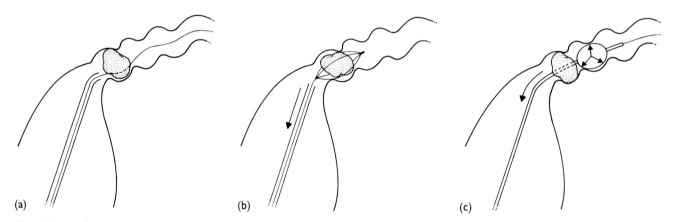

Fig. 5.8 *Methods of removing impacted cystic duct stones*
(a) Using a curved-tipped catheter, a guidewire is negotiated past the stone in the cystic duct. (b) The basket sheath is inserted past the stone over the guidewire, which is then withdrawn. The basket is inserted, the stone engaged and removed. (c) Alternatively, an occlusion balloon may be inserted past the stone over the guidewire. Inflating the balloon and withdrawing disimpacts the cystic duct stone, which can then be removed with a stone basket.

AFTERCARE

After each stone removal session, an appropriately sized self-retaining catheter is inserted into the gall bladder to maintain access and provide drainage (e.g. 14–20 Fr Foley catheters). Complete stone clearance is confirmed by direct visualization and by contrast cholecystography. The catheter is removed if the cystic duct is patent and at least 4 weeks have elapsed since cholecystostomy insertion to allow fibrous tract formation. The gall bladder should be emptied before catheter removal.

RESULTS AND COMPLICATIONS

The success rate of percutaneous cholecystolithotomy is reported to be between 80 and 100% (Picus *et al.*, 1989; Chiverton *et al.*, 1990; Cope *et al.*, 1990). The number of cases, however, is relatively small.

The complications are similar to those of percutaneous cholecystostomy. Since a much larger tract is required for cholecystolithotomy, problems include difficulty in dilating the track to the required size due to resistance from the gall bladder wall (13 to 32%) and loss of access during either dilation or stone extraction. Fortunately, most reports of loss of access during percutaneous cholecystolithotomy have not resulted in bile peritonitis. Other reported complications include dislodgement of suture anchors and inability to remove them after the procedure. Wound infection and vasovagal episodes have also been reported.

Methyl tert-butyl ether

Methyl *tert*-butyl ether is a potent cholesterol solvent. Since 80–90% of gall stones consist predominantly of cholesterol, MTBE dissolution appears to be a suitable method of treatment. However, MTBE is flammable, has a detectable odour, and induces anaesthesia in the patient and surrounding medical personnel. It also causes duodenitis on direct contact and systemic absorption may cause intravascular haemolysis. Therefore, extreme care is required when it is used and the suite where the ether is infused must be specially designed to be spark free and have adequate ventilation to remove any fumes (Williams *et al.*, 1990).

INDICATIONS AND CONTRAINDICATIONS

Gall stones containing more than 40% cholesterol are generally suitable for dissolution (Allen *et al.*, 1985). However, the presence of calcium bilirubinate, other contaminants or rim calcification around the stone adversely affect the ability of MTBE to dissolve the stone. Computed tomography (CT) is particularly useful in predicting the solubility of gall stones. Studies have shown a consistent inverse relationship between the cholesterol content of stones and their CT density (Brakel *et al.*, 1990); the presence of rim calcification has also been accurately detected. Thus, high-resolution thin-slice CT should be performed in patients before MTBE therapy. The presence of any calcification usually means that dissolution will be unsuccessful.

Patients with acute liver, pancreatic or gall bladder disease are not suitable for the procedure. An oral cholecystogram should be performed before dissolution therapy to confirm that the cystic duct is patent and that the gall bladder is functioning. Since a percutaneous cholecystostomy is inserted for MTBE infusion, contraindications to this procedure include those for percutaneous cholecystostomy.

TECHNIQUE

A percutaneous cholecystostomy is performed usually by the transhepatic rather than the subcostal approach. The transhepatic route is preferred because there is, theoretically, less risk of bile leakage around the gall bladder puncture. After the puncture, the cholecystostomy catheter is inserted and looped within the gall bladder body. Catheters used are of a small calibre (e.g. 5 Fr) and have multiple side holes. Once the catheter is in place, the volume of the gall bladder is assessed by first completely emptying it of bile and then replacing it with contrast medium until overflow into the cystic duct is observed fluoroscopically. The ideal volume of MTBE should be less than the volume that causes overflow but should be sufficient to surround the gall stones. The volume usually ranges from 3 to 10 ml.

Once the catheter is secured, the patient is transferred to the 'ether room'. The measured volume of MTBE is then continuously injected and aspirated about five times a minute using a glass syringe (MTBE dissolves plastic syringes) (Fig. 5.9). The MTBE solution in the syringe is changed every 5–15 min to prevent saturation with cholesterol. The progress of dissolution is assessed by performing a contrast cholecystogram through the catheter every 3–4 hours. After all the stones have dissolved, the procedure is continued for a further 2 hours in case small fragments remain. The system is flushed with 50–100 ml saline and the catheter is then removed.

Because this procedure is highly labour intensive, special pumps have been designed to deliver the MTBE solution automatically (McCullough *et al.*, 1989).

Infusion times range from 3 to 30 hours (average 12 hours) and the infusion often takes place over the course of several days. Dissolution may be enhanced by increasing the contact between MTBE and the stones. Thus, bile should be continuously removed from the gall bladder during dissolution and the

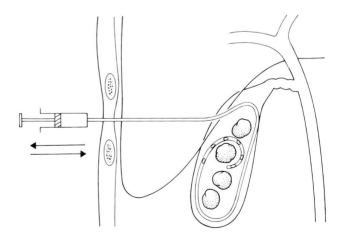

Fig. 5.9
Methyl *tert*-butyl ether (MTBE) dissolution. Standard transhepatic cholecystostomy is performed. A 5 Fr pigtail catheter with multiple side holes is inserted and coiled in the gall bladder. The MTBE is alternatively injected and aspirated.

stones should be agitated by forcefully injecting and aspirating MTBE.

RESULTS AND COMPLICATIONS

The results of dissolution therapy depend on the composition of the gall stones. Success rates of 90% have been reported (Williams *et al.*, 1990). Research is currently underway to assess the effect of using ESWL, mechanical lithotripsy before dissolution of even calcified stones (Lu *et al.*, 1990) and external sonication during dissolution (Griffith *et al.*, 1990; Lu *et al.*, 1992).

The complications of MTBE dissolution include those for percutaneous cholecystostomy insertion (see above). In particular, dislodgement of the MTBE catheter may occur because of its small size and flexible nature. Localized leakage of MTBE around the entry site of the catheter into the gall bladder, manifested by abdominal pain, has been reported. This should be managed by postponing further infusion for 24 hours and reducing the MTBE infusion volume. Minor complications of nausea and vomiting have been observed during therapy. Rarely, systemic effects of MTBE (sedation, intravascular haemolysis and duodenitis) have been reported.

The rotary gall stone lithotrite

The Kensey–Nash lithotrite, modified for use in the gall bladder from the arterial Kensey–Nash catheter used for atherectomy, is currently undergoing clinical evaluation. The device directly fragments gall stones using an impeller that rotates at up to 40000 rpm. As the impeller rotates, a vortex is formed that pulls gall stones against it, fragmenting them (Fig. 5.10). The impeller is surrounded by a six-pronged protective cage that prevents the gall bladder wall from being sucked in simultaneously. The lithotrite has been shown to fragment stones to less than 500 μg, theoretically allowing any retained fragments to pass unimpeded through the cystic duct. Several articles are available on the subject (Miller *et al.*, 1989; Miller and Rose, 1990).

INDICATIONS

All types of stones (i.e. cholesterol, pigment and calcified) are suitable for fragmentation. Certain criteria, however, need to be fulfilled for its use.

The volume of stones should not exceed 75% of the gall bladder volume because of lack of space for

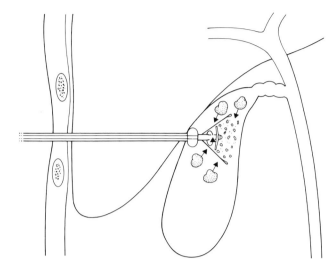

Fig. 5.10
Kensey–Nash lithotrite. Standard transhepatic cholecystostomy is performed. The lithotrite is inserted through a 9.5 Fr stainless steel sheath. Rotation of the impeller forms a vortex, pulling stones against it. A six-pronged protective cage around the impeller prevents gall bladder mucosa from being sucked in.

the device. Also, the use of the lithotrite is currently restricted to stones less than 25 mm.

PATIENT PREPARATION

Epidural anaesthesia or intravenous opiates and benzodiazepines are used for analgesia and sedation. Broad-spectrum antibiotics are given before the procedure.

TECHNIQUE

A transhepatic cholecystostomy is performed and the track is serially dilated to allow insertion of a 9.5 Fr stainless steel sheath. The bile in the gall bladder is completely aspirated and replaced with dilute contrast medium to outline the stones. The patient is placed in a partly upright position (20–30°) so that stones congregate in the body of the gall bladder. The lithotrite is inserted and positioned using fluoroscopic guidance. The device is then activated and the speed of the impeller is gradually increased to 30000–40000 rpm. The procedure is terminated once no fragments are seen fluoroscopically. Procedure time usually ranges from 5 to 30 min. After the procedure, a 10 Fr pigtail catheter is left in the gall bladder and gentle irrigation with saline is performed to remove residual debris. Irrigation is continued intermittently over the next few days. The cholecystostomy catheter is removed after 1 week if the cystic duct is patent and no stone fragments are visible on catheter cholecystography.

RESULTS, COMPLICATIONS AND LIMITATIONS

Reports on the long-term results of this new procedure are still awaited. Potential complications include those for percutaneous cholecystostomy. Complications specific to the technique include gall bladder wall injury due to damage from stone fragments. A limitation of the procedure is that only mobile stones are fragmented. Adherent or impacted cystic duct stones are not affected.

REFERENCES

ALLEN MJ, BORODY TJ, BUGLIOSI TF, MAY GR, LaRUSSO NF, THISTLE JL (1985). Cholelitholysis using methyl tertiary butyl ether. *Gastroenterology* **88**: 122–5.

BECKER CD, QUENVILLE NF, BURHENNE HJ (1988). Long-term occlusion of the porcine cystic duct by means of endoluminal radiofrequency electrocoagulation. *Radiology* **167**: 63–8.

BOGAN ML, HAWES RH, KOPECKY KK, GOULET RJ JR (1990). Percutaneous cholecystolithotomy with endoscopic lithotripsy by using a pulsed-dye laser: Preliminary experience. *American Journal of Roentgenology* **155**: 781–4.

BRAKEL K, LÁMERIS JS, NIJS HGT, TERPSTRA OT, STEEN G, BLIJENBERG BC (1990). Predicting gallstone composition with CT: *In vivo* and *in vitro* analysis. *Radiology* **174**: 337–41.

CESARANI F, GANDINI G, RIGHI D, JULIANI E, RECCHIA S, FRONDA GR (1988). High frequency water jet in the treatment of bile duct lithiasis. *Radiologica Medica* **76**: 453–7.

CHIVERTON SG, INGLES JA, HUDD C, KELLETT MJ, RUSSELL RCG, WICKHAM JEA (1990). Percutaneous cholecystolithotomy: The first 60 patients. *British Medical Journal* **300**: 1310–12.

COPE C (1988). Percutaneous subhepatic cholecystostomy with removable anchor. *American Journal of Roentgenology* **151**: 1129–32.

COPE C (1990). Novel nitinol basket instrument for percutaneous cholecystolithotomy. *American Journal of Roentgenology* **155**: 515–16.

COPE C, BURKE DR, MERANZE SG (1990). Percutaneous extraction of gallstones in 20 patients. *Radiology* **176**: 19–24.

FINNEGAN PW, YEUNG EY, THURSTON W, HO C-S (1991). Outpatient radiologic percutaneous cholecystolithotomy: preliminary results. Presented at Canadian Association of Radiologists (Annual Meeting).

GIBNEY RG, CHOW K, SO CB, ROLEY VA, COOPERBERG PL, BURHENNE HJ (1989). Gallstone recurrence after cholecystolithotomy. *American Journal of Roentgenology* **153**: 287–9.

GRIFFITH SL, BURNEY BT, FRY FJ, FRANKLIN TD (1990). Experimental gallstone dissolution with methyl *tert*-butyl

ether (MTBE) and transcutaneous ultrasound energy. *Investigative Radiology* **25**: 146–52.

HO C-S, YEE AC, McLOUGHLIN MJ (1987). Biliary lithotripsy with a mechanical lithotripter. *Radiology* **165**: 791–3.

HWANG MH, MO LR, CHEN GD, YANG JC, LIN CS, YUEH SK (1987). Percutaneous transhepatic cholecystic ultrasonic lithotripsy. *Gastrointestinal Endoscopy* **33**: 301–3.

LARSSEN TB, GÖTHLIN JH, JENSEN D, ARNESJÖ B, SORIEDE Ø (1988). Ultrasonically and fluoroscopically guided therapeutic percutaneous catheter drainage of the gallbladder. *Gastrointestinal Radiology* **13**: 37–40.

LU DSK, HO C-S, ALLEN LC (1990). Gallstone dissolution in methyl *tert*-butyl ether after mechanical fragmentation: *In vitro* study. *American Journal of Roentgenology* **155**: 67–72.

LU DSK, HO C-S, FOSTER FS, HUNT JW, YEUNG E, RICKETT J, McKINLEY C, ALLEN LC (1992). Accelerated gallstone dissolution in methyl *tert*-butyl ether by sonication: *In vitro* study *Investigative Radiology* **27**(5); 356–61.

McCULLOUGH JE, LESMA A, THISTLE JL (1989). A rapid stirring automatic pump system for dissolving gallstones using methyl *tert*-butyl ether (MTBE): *In vitro* comparison with the manual syringe method. *Gastroenterology* **96**: a629 (abs).

McSHERRY CK (1981). The national cooperative gallstone study report: A surgeon's perspective. *Annals of Internal Medicine* **95**: 379.

MILLER FJ, KENSEY KR, NASH JE (1989). Experimental percutaneous gallstone lithotripsy: Results in swine. *Radiology* **170**: 985–7.

MILLER FJ, ROSE SC (1990 Intervention for gallbladder disease. *Cardiovascular and Interventional Radiology* **13**: 264–71.

PEARSE DM, HAWKINS IF, SHAVER R, VOGEL S (1984). Percutaneous cholecystostomy in acute cholecystitis and common duct obstruction. *Radiology* **152**: 365–7.

PICUS D, MARX MV, HICKS ME, LANG EV, EDMUNDOWICZ SA (1989). Percutaneous cholecystolithotomy: Preliminary experience and technical considerations. *Radiology* **173**: 487–91.

RAWAT B, BURHENNE HJ (1990). Biliary extracorporeal shock-wave lithotripsy. *Cardiovascular and Interventional Radiology* **13**: 258–63.

TEPLICK SK, BRANDON JC, WOLFERTH CC, AMRON G, GAMBESCIA R, ZITOMER N (1990). Percutaneous interventional gallbladder procedures: Personal experience and literature review. *Gastrointestinal Radiology* **15**: 133–6.

VAN SONNENBERG E, WING VW, POLLARD JW, CASOLA G (1984). Life-threatening vagal reactions associated with percutaneous cholecystostomy. *Radiology* **151**: 377–80.

VAN SONNENBERG E, WITTICH GR, CASOLA G, PRINCENTHAL RA, HOFMANN AF, KEIGHTLEY A, WING VW (1986). Diagnostic and therapeutic percutaneous gallbladder procedures. *Radiology* **160**: 23–6.

VOGELZANG RL, NEMCEK AA JR (1988). Percutaneous cholecystostomy: Diagnostic and therapeutic efficacy. *Radiology* **168**: 29–34.

WILLIAMS HJ, BENDER CE, LEROY AJ (1990). Dissolution of cholesterol gallstones using methyl *tert*-butyl ether. *Cardiovascular and Interventional Radiology* **13**: 272–7.

YOSHIMOTO H, IKEDA S, TANAKA M, MATSUMOTO S, KURODA Y (1989). Choledochoscopic electrohydraulic lithotripsy and lithotomy for stones in the common bile duct, intrahepatic ducts, and gallbladder. *Annals of Surgery* **210:** 576–82.

I would like to thank Ms Edvige Coretti for her expert secretarial assistance in the preparation of this manuscript.

CHAPTER 6

Endoscopic and combined management of biliary strictures

Antony G. Speer and Robert N. Gibson

Epidemiology and risk factors   82

Endoscopic or percutaneous stenting?   82

Prestenting assessment   83

Technique   83

Results   90

Combined percutaneous endoscopic stenting   92

Benign strictures   93

Conclusions   95

References   96

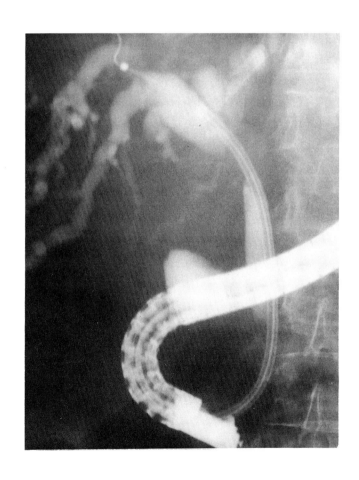

EPIDEMIOLOGY AND RISK FACTORS

Patient selection is the key to obtaining good results in the management of malignant obstructive jaundice. A broad understanding of tumour biology, epidemiology and risk factor analysis is helpful in selecting the best technique for each patient.

The majority of malignant obstructive jaundice is caused by carcinoma of the pancreas, carcinoma of the gall bladder, cholangiocarcinoma or metastases. Epidemiological studies have shown that carcinoma of the pancreas is four to five times more common than carcinoma of the gall bladder or cholangiocarcinoma. The median age at presentation of all three primary tumours is approximately 70 years (Fraumeni, 1975). That is, half the patients with malignant obstructive jaundice are older than 70. Carcinoma of the pancreas is a biologically aggressive tumour, it spreads early and presents late. Only a small proportion of patients are suitable for resection with the possibility of a cure. The outlook for cholangiocarcinoma and carcinoma of the gall bladder is similarly poor.

The role of various risk factors has been studied in detail for surgery in carcinoma of the pancreas. The risk of biliary surgery increases with increasing age, with a sharp increase above 70 years of age, mainly due to complications of coexisting diseases such as ischaemic heart disease and chronic obstructive airways disease. The risk of surgery also increases with the extent of metastatic disease. Operative mortality is approximately double in the presence of liver metastases and doubles again if there is malignant ascites. Other risk factors such as impaired renal function, low albumin and raised white cell count have also been identified.

Intuitively, one feels that these factors should also apply in cholangiocarcinoma and carcinoma of the gall bladder, although this has not been extensively studied.

The other important prognostic feature in hilar lesions is the extent of intrahepatic stricturing. All techniques have difficulty with multiple intrahepatic strictures.

ENDOSCOPIC OR PERCUTANEOUS STENTING?

Surgical palliation of elderly, frail patients has a high morbidity and mortality. In the late 1970s minimally invasive stenting techniques were developed for this high-risk group. Initially it was not clear which technique was the best. Percutaneous stenting seemed technically easier, however, the endoscopic route did not require a track through the liver. In the early 1980s the two techniques were compared in a prospective randomized trial at the Middlesex and London Hospitals (Speer *et al.*, 1987). Elderly, frail patients who were high risk for surgery and had biliary obstruction due to unresectable malignancy were randomized to receive either a percutaneous or endoscopic stent. The procedures were performed by experienced gastroenterologists and interventional radiologists who previously had published series with results equal to those anywhere in the world. The trial found that endoscopic stenting had a significantly lower 30-day mortality and this was due to the higher incidence of complications in the percutaneous group. These complications were mainly bile leakage and haemorrhage due to puncturing of the liver. The group concluded that endoscopic stenting was the preferred route for palliation of these elderly, frail patients and should be attempted first.

Not surprisingly, this trial has been widely debated, but never repeated. The general consensus from units with broad experience in percutaneous, endoscopic and surgical techniques is that for low lesions there is no doubt that endoscopic stenting is best. Apart from the lower morbidity and mortality, there is much less patient discomfort. For hilar lesions the situation is not quite so clear. The percutaneous technique has the advantage of being able to select the best segment of liver to drain. This is not always possible endoscopically. Furthermore, expanding metal stents can be inserted down a relatively small diameter transhepatic track, possibly reducing complications. Our preference is to perform endoscopic stenting if a duct draining a substantial segment of nonatrophic and tumour-free liver can be entered (see Chapter 2). If this is not possible, then the stricture should be stented percutaneously or possibly at a combined percutaneous–endoscopic ('rendezvous') procedure.

Prestenting assessment

Patients should be carefully reviewed to identify any coexisting medical problems. Smoking is one of the few factors identified as a cause of carcinoma of the pancreas and many patients have coexisting chest or cardiovascular problems. Obviously if these medical problems would prevent major surgery such as Whipple's resection, then there is no need to perform detailed imaging to assess resectability.

It is beyond the scope of this chapter to discuss imaging in detail. Assessment is usually performed with a combination of computed tomography (CT), ultrasound and cholangiography (see Chapter 2). Cholangiography is preferably performed immediately before any planned intervention to reduce the risk of cholangitis or bile leak.

Imaging should be planned to answer the following questions:

1 What is the level of obstruction?
2 Is the lesion resectable? Evidence of distant or local invasion is important for all levels of stricture. For hilar strictures, it is necessary to determine the extent of intrahepatic stricturing, in particular, whether or not the left and right ducts communicate, and to identify any atrophic lobes. The best segment of liver to be drained should be clearly identified.

The 'no intervention' option

One of the most difficult decisions to make is when not to intervene. There are two broad groups of patients where this may be considered: those who are very sick and those who have few symptoms due to biliary obstruction.

In elderly, frail patients with extensive metastases, intervention may cause more distress than palliation. Life expectancy is often very short and pruritus may be reasonably well controlled with medication. On the other hand, tumours such as cholangiocarcinoma may grow very slowly causing few symptoms initially. Figure 6.1 shows the cholangiogram of a patient who presented with abnormal liver function tests. Strictures extended into the second-order ducts on the right and into the left hepatic duct. An abdominal CT scan confirmed an atrophic left lobe precluding resection by right hepatectomy. Her bilirubin was normal, she had no pruritus and remained well for 18 months with conservative management. Figure 6.2 shows the cholangiogram of a patient who presented with mild jaundice. There is

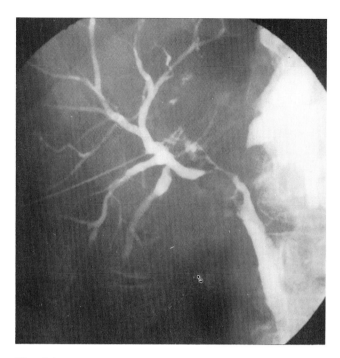

Fig. 6.1
Percutaneous transhepatic cholangiogram showing a hilar stricture involving the left hepatic duct and the second-order right hepatic ducts. The left lobe is atrophied. The patient had no symptoms and was managed conservatively.

extensive stricturing of the left and right hepatic ducts, however, an anomalous right anterior duct inserting low into the common hepatic duct was not initially involved. This patient was managed conservatively for 8 months at which time she developed increasing jaundice and a stent was placed in the right anterior duct endoscopically.

Technique

Patient preparation

Once a decision has been made to recommend stenting then consent should be obtained. The endoscopist is ultimately responsible for ensuring the patient makes a well-informed decision. The procedure is described, preferably with the aid of diagrams and written explanations. Possible complications are mentioned as well as the risk of death. This part of the discussion should be tailored to the level of obstruction; low lesions have a small incidence of complications, whereas, hilar strictures, particularly those with multiple intrahepatic stric-

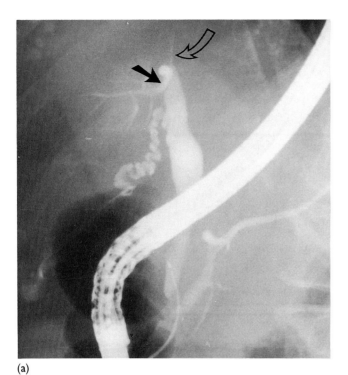
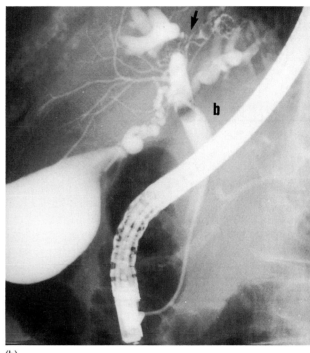

(a)   (b)

Fig. 6.2
(a) Endoscopic retrograde cholangiogram showing a hilar stricture with an anomalous right sectoral duct (solid arrow) joining the common hepatic duct below the stricture (open arrow). (b) A balloon catheter 'b' has been passed above the cystic duct and inflated to prevent overfilling of the gall bladder. The hilar stricture (arrow) extends into the right and left hepatic ducts. The right lobe, drained by the anomalous right sectoral duct, is hypertrophied.

tures, have a substantial morbidity and mortality. The treatment options should be discussed.

The patient's clotting should be checked and corrected with intramuscular vitamin K or clotting factors if required. Immediately before the procedure, if the patient is fasting for a considerable time, consideration should be given to maintaining hydration with intravenous fluids. Elderly, jaundiced patients quickly develop renal impairment with dehydration.

PROPHYLACTIC ANTIBIOTICS

Antibiotics should be given prophylactically 1–2 hours before the procedure to obtain adequate tissue levels during stenting. Our preference is to use a ureidopenicillin such as piperacillin. This has the advantage of a broad spectrum of activity, low toxicity and achieves excellent levels in bile once the obstruction is relieved (Dooley *et al.*, 1984).

FACILITIES

The procedure should be performed in a fluoroscopy room similar to that required for percutaneous stent-

ing. Experienced nurses should be available to help with the equipment and monitor the patient.

SEDATION

Ideally, an anaesthetist should be available for the procedure. This is not always possible in a busy hospital. The authors' preference is to use a combination of intravenous midazolam and pethidine with the dose titrated according to the patient's body weight, age and observed effect. Hyoscine-*N*-butylbromide (Buscopan) is used to depress duodenal motility. This may cause tachycardias in the elderly and those with cardiovascular problems. Glucagon should be used in these situations.

The patient's condition should be monitored closely by an experienced nurse throughout the procedure. Pulse oximetry should be used in all patients and an automatic pulse and blood pressure machine may be considered for those who are at risk.

Equipment and technique

Purpose-built stenting duodenoscopes are made by a variety of companies. These are slightly longer than a conventional endoscope, the endoscopic view is to the side rather than forward and there is a wide instrument channel (3.8 to 4.2 mm in diameter). Both fibreoptic and video instruments are available. The catheters and guidewires are similar to those used in radiology, but they are usually much longer (200–300 cm) and are not so torque stable.

After the patient is adequately sedated the endoscope is passed to the duodenum. A 6 Fr catheter is inserted in the papilla, a cholangiogram is obtained and the position and extent of the stricture defined (Fig. 6.3a). The catheter is then manoeuvred deep into the bile duct. A small sphincterotomy is required in some cases to facilitate this. An 0.035 inch floppy tipped guidewire is passed down the catheter and manoeuvred through the stricture (Fig. 6.3b). If the duct entered is not suitable for stenting the wire should be repositioned into a suitable duct (Fig. 6.3c). The catheter is then pushed through the stricture over the guidewire. Further contrast medium can be injected above the stricture to outline the ducts being drained (Fig. 6.3d). Bile should be aspirated for cytology and culture. A wire-guided endoscopy brush may also be used to take cytological specimens directly from the stricture. In low strictures due to carcinoma of the pancreas or those due to external compression, a stent can be inserted directly over the catheter without further dilation. Hilar strictures are often tight and tortuous and it is prudent to dilate these with a coaxial dilator passed over the catheter or guidewire before stent insertion (Fig. 6.3e). After dilation the stent is pushed down through the endoscope and up through the stricture with a coaxial 10 Fr pushing catheter. The lower 1–2 cm of the stent is left in the duodenum to assist stent replacement (Fig. 6.3f). Both 10 and 11.5 Fr polyethylene stents are used. These are shaped to the contour of the bile duct. A variety of lengths are available according to the site of the stricture.

DIFFICULT SITUATIONS

Deep cannulation of the bile duct is difficult if there is tumour invading or distorting the papilla. A precut sphincterotomy with a needle knife gives immediate access in 50–60% of such patients. A further 20% can be cannulated after 24–48 hours. If this is unsuccessful, a combined percutaneous–endoscopic approach could be used.

Hilar strictures are often tight, fibrous and difficult to dilate. If the inner catheter cannot be passed through the stricture, then stiff graduated dilators can be tried. If these are unsuccessful, a small diameter nasobiliary drain can be used to decompress the duct. It is then usually possible to dilate the stricture after several days of drainage.

In some tight hilar strictures it may be difficult to pass a guidewire. The catheter should be placed just below the stricture and radiographs taken while injecting contrast medium. These usually show the position and course that the stricture takes from the normal duct. A variety of guidewires can be tried, including moveable core, J-tip and wires coated with hydrogel (Radifocus wire, Terumo).

Assessment of stent function

After a stent has been inserted the serum bilirubin should be measured daily for several days. The level will fall steadily and often dramatically in low strictures. In patients with hilar strictures, particularly when only a portion of the liver is drained, the serum bilirubin will fall slowly and may even rise a little initially. Pre-existing chronic liver disease or an increased bilirubin load from a bleeding sphincterotomy, erosions or peptic ulcer may also increase the bilirubin level or delay the fall. If the bilirubin level is falling then the stent is functioning adequately and no further investigations are needed. If the level is static for several days or rising, the stent may not be draining.

This problem is best investigated with abdominal ultrasound. The two questions to answer are:

1 Is the stent in a good position across the stricture?
2 Are the ducts drained by the stent decompressed?

On ultrasound the stent can be easily seen within the bile duct and the position of the upper end confirmed. If the stent is draining adequately, the ducts above it will be decompressed compared with previous images and may contain air.

If the examination shows that the stent is in a good position with the ducts above it decompressed then the rising bilirubin level may be due to one of the factors discussed above and the patient could be managed conservatively. However, if the stent has migrated out of the stricture, the ducts are not decompressed or the imaging is equivocal, then the stent should be replaced.

During long-term follow-up blocked stents present with a recurrence of jaundice or the onset of cholangitis. This does not usually present a diagnostic problem. However, minor elevations of bilirubin or deterioration in liver function may be due to early stent blockage, liver metastases or even

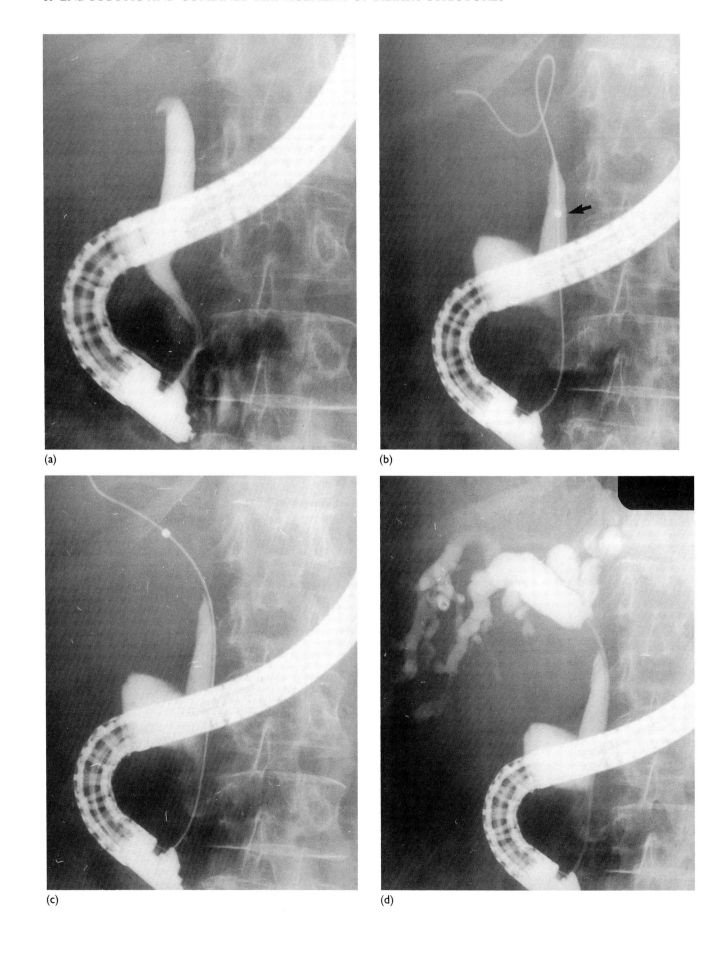

(a)

(b)

(c)

(d)

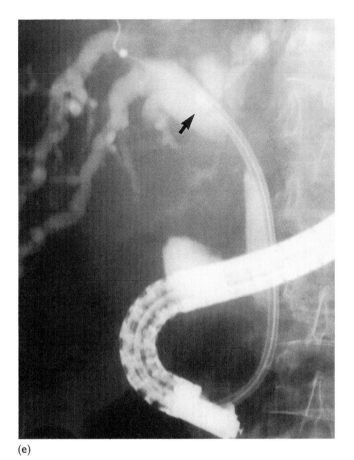

(e)

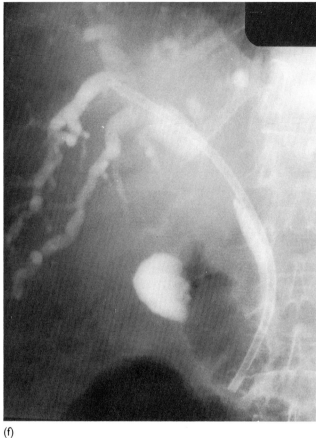

(f)

Fig. 6.3
(a) Endoscopic retrograde cholangiogram showing a hilar stricture. (b) A 6 Fr catheter, with a metal ring to mark the tip (arrow) is deep in the duct. A guidewire has been passed through the stricture but there is a tight loop in the duct selected, making it unfavourable for stenting. (c) The guidewire has been repositioned in a duct more suitable for stenting and the catheter has been pushed through the stricture over the wire. (d) The guidewire has been removed and more contrast medium injected to outline the ducts above the stricture. (e) An 11.5 Fr dilating catheter has been passed over the inner catheter and through the stricture. A metal ring (arrow) marks the end of the taper on the dilator. (f) A 10 Fr stent is in place across the stricture. The lower 2 cm of the stent remains in the duodenum.

portal vein invasion. Ultrasound is the investigation of choice to distinguish between these diagnostic possibilities.

Assessment of fever and/or right upper quadrant pain poststenting

Fever with or without associated right upper quadrant discomfort is a common problem occurring after both endoscopic and percutaneous stenting. Mild fever that resolves with antibiotics requires no further investigation and is probably due to transient obstruction of the stent. If the patient has fever despite receiving antibiotics or the fever recurs when the antibiotics are stopped, then the diagnosis needs to be pursued further. The most common cause is a

poorly draining stent, and this can be assessed with ultrasound as discussed previously.

The presence of intrahepatic abscesses can usually be determined by ultrasound although CT may also be needed. Large abscesses will generally require percutaneous drainage, but small abscesses, which may be multiple, are best managed by appropriate antibiotic therapy and ensuring that there is good biliary drainage of that lobe or segment.

Cholecystitis may occur particularly if the cystic duct is obstructed by tumour or if stones are present in the gall bladder. Again ultrasound is the investigation of choice.

Retroperitoneal perforation resulting from the sphincterotomy or duct perforation during stenting are unusual, but should always be considered. The radiographs taken during the stenting procedure should be reviewed for evidence of contrast medium or air in the retroperitoneum. Abdominal CT gives the best delineation of retroperitoneal or intra-abdominal collections.

Infection in an undrained segment can occur in patients with multiple intrahepatic strictures. This should be considered when other possibilities have been excluded. The diagnosis is made by performing a percutaneous transhepatic cholangiogram of the suspected segment and aspirating bile for culture. If the bile is obviously purulent, immediate percutaneous drainage of that segment should be performed.

Management of complications

Cholangitis is best avoided by cleaning instruments thoroughly, using prophylactic antibiotics and obtaining good drainage. If cholangitis occurs in undrained segments in hilar strictures, these usually require percutaneous drainage and vigorous treatment with antibiotics.

Local perforation at the site of sphincterotomy can usually be managed conservatively with intravenous antibiotics and nil by mouth. This complication is best avoided by performing a small sphincterotomy. Perforation can also occur if guidewires or catheters are passed through the wall of the bile duct, usually below the stricture. This often causes few clinical symptoms and again can be managed conservatively with intravenous antibiotics and nil by mouth. Care should be taken when manoeuvring guidewires and catheters, and force should never be used in manipulations.

Haemorrhage from the sphincterotomy is usually minor and stops spontaneously. Major bleeding may respond to endoscopic injection of adrenaline or very occasionally requires surgery or transarterial embolization.

Cholecystitis is an unusual complication and is best avoided by using prophylactic antibiotics. Young, fit patients should be managed conventionally with a cholecystectomy. Elderly, frail patients usually respond to percutaneous cholecystostomy and vigorous antibiotic therapy.

Stent replacement

ENDOSCOPIC STENTS

Stents should be replaced when they are blocked or electively at 3 to 4 months after insertion if the patient remains reasonably well. Endoscopic stent

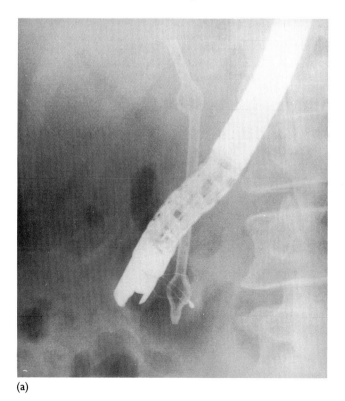

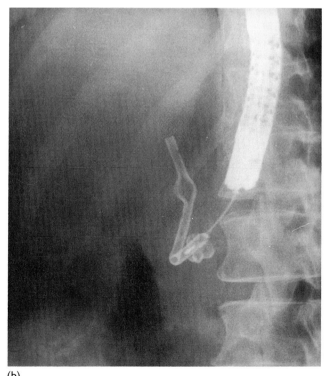

(a)

(b)

Fig. 6.4 *Stent removal*
(a) The lower end of a blocked double-mushroom percutaneous stent has been grasped with an endoscopic basket. (b) The endoscope, basket and double-mushroom percutaneous stent are pulled out through the patient's mouth.

replacement is technically easier than the original procedure and can usually be performed in 20–30 mins. Elective replacements can be performed as day cases, the patient arrives fasting in the morning and returns home after a few hours of observation.

The lower end of an endoscopic stent is left in the duodenum to facilitate replacement. This is grasped with a basket and the endoscope, with the stent locked onto the tip, and is then removed out through the patient's mouth. Figure 6.4 shows the removal of a blocked double-mushroom percutaneous stent. The lower end has been grasped with a basket (Fig. 6.4a) and the endoscope and stent then removed (Fig. 6.4b). The endoscope is then re-inserted and another stent positioned across the stricture in the usual manner.

PERCUTANEOUS STENTS

Percutaneous stents can be replaced as easily as endoscopic stents if the lower end is in the duodenum as is the case with most low strictures. If the lower end is in the bile duct, then stent removal is technically more difficult, but not impossible for an experienced endoscopist. Occluded stents can be replaced by a percutaneous approach but this is technically difficult. In our experience, endoscopic replacement is

much easier and quicker for both the patient and the operator. We would recommend endoscopic replacement as the treatment of choice for all blocked stents.

EXPANDING METAL STENTS

Expanding metal stents are now being inserted by both the percutaneous and endoscopic routes. Despite their large lumen, blockage occurs by a combination of tumour ingrowth and biliary sludge. Tumour growth into the metal framework prevents removal of these stents and a variety of techniques have been described to re-establish drainage. These include destruction of the tissue in the lumen of the stent with a laser or diathermy probe, or insertion of a standard 10 or 12 Fr polymer stent through the occluded lumen. We favour stenting as it is easily performed endoscopically and has few complications. Figure 6.5a shows a metal stent that has been inserted percutaneously through a hilar stricture. The lumen is blocked and the patient has presented with a recurrence of jaundice. An endoscopic stent has been placed through the lumen relieving the obstruction (Fig. 6.5b).

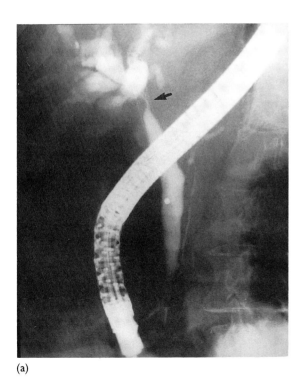

(a)

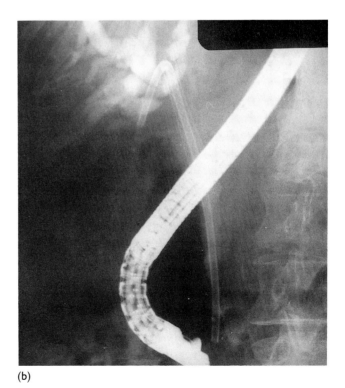

(b)

Fig. 6.5
(a) An expanding metal stent (Wallstent) has been placed percutaneously across a hilar stricture. The patient presented with jaundice and endoscopic cholangiography shows that the lumen of the stent is narrowed (arrow). (b) A 10 Fr stent has been placed through the lumen of the blocked expanding metal stent to relieve the obstruction.

RESULTS

When considering the results of published series using different techniques, care should be taken to pay due attention to the confounding variables of patient selection and the differing definitions of complications. In general, with all techniques and particularly with surgery, elderly, frail patients have a higher morbidity and mortality. The definitions of complications such as cholangitis or haemorrhage differ widely. Some series include only those complications which require several days in an intensive care unit, whereas other prospective randomized trials have strict criteria with prospective evaluation of each patient resulting in a spuriously high incidence of complications. The term 'procedure-related mortality' also suffers from potential observer bias and depends on the diligence of observation and documentation of complications, in particular, at *post mortem* examination. The best measure of early morbidity for interseries comparison is 30-day mortality. This is not subject to observer bias and is truly comparable between series, depending mainly on the early complications of the technique.

Low biliary strictures

The results of endoscopic stenting are best considered in three different risk categories for surgery: low, medium and high. The risk of surgery increases with increasing age, increasing metastatic disease and with associated medical conditions.

The last two variables are not always clearly described in series and it is convenient to identify these risk categories simply by their average age: low risk in those whose age is in the low 60s, medium risk at around 70 years of age, and high risk in those whose age is in the high 70s.

High-risk patients

The results of stenting in high-risk patients with carcinoma of the pancreas are summarized in Table 6.1. Endoscopic stents can be successfully inserted in about 90% of patients with relief of jaundice. The 30-day mortality is 9–10% and the mean survival time is 6 months. In these two series, 20–30% of patients developed stent blockages requiring replacement. This was performed endoscopically requiring a short admission. More recently, endoscopists have tended to change stents electively after 3–4 months, avoiding the potential complication of stent blockage. The incidence of late duodenal

Table 6.1 *Results of endoscopic stenting for carcinoma of the pancreas in patients who are high risks for surgery.*

	University of Amsterdam (Huibregtse et al., 1987)	The Middlesex Hospital (Speer & Cotton 1988)
n	221	99
Age (median years)	71	76
Relief of Jaundice	87%	88%
30 Day Mortality	10%	9%
Mean Survival (weeks)	26	25
Surgery for duodenal obstruction	7.5%	6%
Stent change	21%	29.5%

obstruction is only 7% which, because of patient selection, is lower than would be expected from surgical series. Duodenal compression or invasion makes stenting technically difficult and these patients are best managed with a gastrojejunostomy as well as a biliary-enteric anastomosis (so-called 'double bypass').

Medium-risk patients

As endoscopic stenting developed and was used more widely, consideration was given to palliating patients who had unresectable tumour, but were fit for bypass surgery. In medium-risk patients stenting has been compared with bypass surgery in two prospective randomized trials, the results of which are summarized in Table 6.2. The 30-day mortality for stenting is approximately half that of surgery and this is due to the lower incidence of complications. The overall survival with the two techniques is similar and depends largely on the tumour itself. The early advantages of stenting are somewhat balanced by the incidence of stent blockage requiring replacement. Elective stent replacement was not practised when these trials were designed. The development of new stents and the use of elective replacement may reduce this long-term disadvantage for stenting.

Low-risk patients

Young, fit patients with small tumours deserve the benefit of surgery with a possibility of resection and cure. The main role for endoscopy is to identify these tumours. There are theoretical considerations that suggest that preoperative drainage may reduce surgical morbidity and mortality, however, prospective randomized trials of percutaneous trans-

Table 6.2 *Randomised trials comparing endoscopic stenting with bypass surgery in unresectable carcinoma of the pancreas in patients who are medium risk for surgery.*

	Wessex Shepherd et al., 1988		Middlesex Smith et al., 1990	
	Stent	Surgery	Stent	Surgery
n	23	25	101	103
Age (median years)	73	73	70	70
Relief of Jaundice (%)	91	92	90	91
30 Day Mortality (%)	9	20	5	19
Survival (Weeks)	22	18	20	24
Surgery for duodenal obstruction (%)	9	4	6	2
Total Hospital Stay Initial & Readmissions (Days)	8	13*	9	13*
Stent change (%)	33		17	

*P < 0.05

hepatic drainage have not confirmed this. Two trials of preoperative endoscopic drainage are currently being performed but the accrual rate is slow. Preoperative endoscopic drainage could be considered in patients presenting with deep jaundice and renal impairment or cholangitis not responding to medical treatment.

AMPULLARY CARCINOMA

Ampullary carcinoma has a much better prognosis than carcinoma of the pancreas. The tumour is slow growing and metastasizes late. The 5-year survival after surgical resection is up to 40%. Young, fit patients should be considered for Whipple's resection while the elderly and frail can be managed with a local wedge resection.

The endoscopist's main task is to diagnose these tumours and obtain histological confirmation. Endoscopic retrograde cholangiopancreatography (ERCP) allows direct visualization of the papilla. Many lesions are obvious with friable or ulcerated tumour replacing the papilla. However, about 30% are small and submucosal. These should be suspected if the papilla is enlarged or the pancreatic and/or bile ducts are dilated down to an irregular stricture at the lower end. If a tumour is suspected

but no obvious malignant tissue can be seen and biopsied, then a sphincterotomy should be performed and brushings and biopsies taken from the cut edge. If these show no evidence of malignancy then biopsies and brushings should be repeated in 2 week's time once the inflammation and oedema from the sphincterotomy has subsided.

There is a small group of patients with ampullary carcinoma who present with metastatic disease or who are risks for surgery due to advanced age or other medical conditions. These patients can be palliated with endoscopic techniques. Huibregtse et al. (1987) have the largest experience and recommend stenting without a sphincterotomy. These tumours are friable and may bleed with a sphincterotomy. In a series of 71 patients Huibregtse successfully stented 98% with relief of jaundice. The 30-day mortality was 2% and the median survival 58 weeks. The major late complication was duodenal stenosis occurring in 23% of cases. Others have palliated these tumours with endoscopic sphincterotomy alone or in combination with a debulking therapy such as laser ablation or diathermy resection.

Hilar strictures

Hilar strictures occur less frequently making the design of randomized trials in this area difficult. These tumours often spread widely along the bile ducts and may be multifocal so surgical resection is technically difficult. However, if a lesion appears resectable on diagnostic imaging and the patient is fit, then we would recommend resection as it gives good palliation with a small chance of a cure. In those who are unresectable or high risks for surgery, endoscopic stenting provides good palliation. The results of representative series are summarized in Table 6.3.

In strictures where both the left and right hepatic ducts communicate, a stent can be inserted in over 90% of cases and will relieve jaundice. However, in those patients with multiple intrahepatic strictures, a stent can only be successfully inserted in about 70% of cases and does not always relieve the jaundice.

In this difficult group of patients there is debate as to whether a single stent or multiple stents to drain multiple segments should be used. Deviere et al. (1988) claim a lower incidence of cholangitis when multiple stents are used, but this finding was not confirmed by Huibregtse (1988). The Middlesex group used a single stent plus vigorous prophylactic antibiotics and had only a modest incidence of cholangitis (Speer and Cotton, 1989). With multiple intrahepatic strictures care must be taken to introduce the stent into a large segment of liver that is not

Table 6.3 *Results of endoscopic stenting in malignant hilar strictures in patients who are high risks for surgery.*

	Amsterdam n = 300 Huibregtse 1988	**Middlesex** n = 64 Speer & Cotton 1989	**Brussels** n = 70 Deviere *et al.,* 1988
Age (years)	69	68	72
Stricture involves both left and right hepatic ducts (%)	100	78	71
Successful Stent Insertion (%)	84	81	89 (97)*
Relief of Jaundice (%)	73	69	?
Early Cholangitis (%)	19	11	24
30 Day Mortality (%)	25	13	15
Survival (Weeks) Mean (Median)	25 (12)	22 (16)	21
Stent Changes (%)	29	25	37

* The higher figure for successful stent insertion was achieved by performing combined procedures if endoscopic stenting failed.

atrophied or extensively invaded by tumour. It is not always possible to select the desired segment endoscopically and, in these cases, the endoscopist should seek the help of a radiologist to perform a combined procedure or percutaneous transhepatic stent.

COMBINED PERCUTANEOUS ENDOSCOPIC STENTING

A combined procedure is a rendezvous in the duodenum between the radiologist and the endoscopist. The percutaneous catheter and guidewire are used to provide endoscopic access through difficult strictures or to select an appropriate segment for drainage in patients with multiple intrahepatic strictures. This procedure has two possible advantages over standard percutaneous stenting. A small 6 or 7 Fr diameter catheter can be used for the percutaneous access to the biliary tree, theoretically reducing the incidence of bleeding and bile leakage compared to the larger diameter track required to introduce a 10 or 12 Fr stent percutaneously. With both ends of a guidewire firmly anchored, much force can be applied to a dilator and it is possible to dilate nearly all tight, tortuous strictures. A variety of techniques has been described and each group seems to have its own variation. We will describe our preferred method.

Low strictures

The combined technique is usually used in low strictures where it is not possible for the endoscopist to obtain selective cannulation of the bile duct due to distortion or invasion of the ampulla by tumour. The radiologist inserts a small catheter into the bile duct using standard techniques and a guidewire is manoeuvred through the stricture into the duodenum. This part of the procedure is performed with the patient supine to facilitate percutaneous access. Once the guidewire and catheter are in place in the duodenum the patient is then carefully rolled into a left lateral semiprone position.

A short curved tip on a torque stable catheter (Fig. 2.14, page 24) will facilitate alignment of the percutaneous and endoscopic catheters. The endoscopic catheter is then advanced over 1–2 cm of guidewire protruding from the percutaneous catheter (Fig. 6.6a). This part of the procedure is best performed with a video endoscope or a camera on a fibreoptic endoscope so that both operators can see the catheters in the duodenum. The radiologist then pushes the guidewire 6–10 cm further into the endoscopist's catheter and then, under fluoroscopic control, slowly withdraws his catheter leaving the guidewire *in situ* while the endoscopist pushes his catheter up into the papilla and through the stricture. The percutaneous guidewire holds the two catheters together and guides the endoscopic catheter into the papilla and through the stricture. Once the tip of the endoscopic catheter is well above the stricture, the radiologist withdraws his guidewire separating the two catheters and the endoscopist introduces a guidewire down his catheter and into the intrahepatic ducts and inserts a stent by the usual technique.

A percutaneous drainage catheter is left *in situ* for 24–48 hours and removed if a repeat cholangiogram shows good stent function.

Hilar strictures

Hilar strictures can be stented with a similar percutaneous–endoscopic technique to that used in low strictures. The location of the tip of the endo-

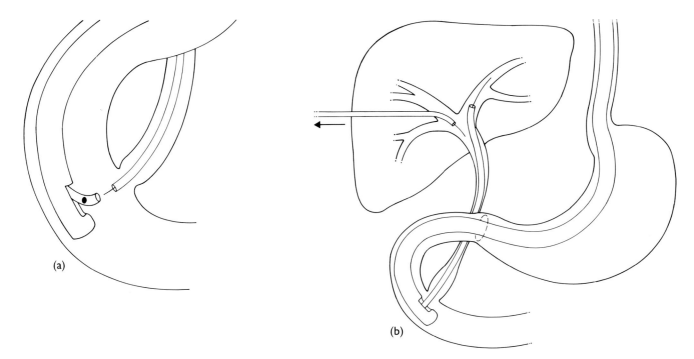

Fig. 6.6 *Diagrams of the combined procedure*
(a) The endoscopic catheter is advanced onto a short length of guidewire protruding from the percutaneous catheter. A torque stable catheter with a curved tip helps to align the two catheters. (b) The endoscopic catheter has been guided through the stricture and the percutaneous guidewire withdrawn, freeing the catheters. The endoscopic guidewire has been placed in the left hepatic duct away from the percutaneous duct puncture site.

scopic catheter above the stricture is more critical in this situation. Care should be taken to avoid introducing the endoscopic catheter into the liver substance along the percutaneous track running the risk of leaving the tip of the stent in the liver parenchyma rather than a bile duct. The exact position of the stricture can be outlined by introducing contrast medium down the endoscopic catheter with the percutaneous wire in place. Once the catheter tip has passed the stricture and the percutaneous wire withdrawn, then the endoscopist should pass the endoscopic guidewire into an intrahepatic duct away from the puncture site (Fig. 6.6b). The radiologist should take care not to lose position in the intrahepatic duct as the catheters separate.

If the stricture is tight and tortuous and it is not possible to dilate it endoscopically, dilation can usually be performed over a guidewire fixed at both ends. A long 400 cm guidewire is passed via the percutaneous catheter and grasped in a basket and pulled out through the endoscope. With both ends secured, a dilator is passed through the endoscope over the guidewire and much more force applied to dilate the stricture. Care must be taken during the procedure to keep the percutaneous guidewire covered by the catheter to prevent 'cheese-wiring' of the liver.

Endoscopists tend to require combined procedures less frequently as they become more experienced and skilled. The role of combined procedures compared with percutaneous stenting with small diameter expanding metal stents remains undecided.

BENIGN STRICTURES

Sclerosing cholangitis

Sclerosing cholangitis occurs infrequently and has a variable clinical course with unpredictable relapses and remissions. Some patients have multiple intrahepatic strictures and if liver failure develops the only feasible form of treatment is liver transplantation. Others have a dominant stricture in the common bile duct or hepatic duct with less extensive intrahepatic strictures, and may benefit from treatment of their dominant stricture.

Surgery is one option, the stricture being excised with a Roux loop anastomosis to the bile duct. The Roux loop is then attached to the anterior abdominal wall and marked with clips (see Chapter 4). This allows easy percutaneous access to the biliary tree for further dilation of intrahepatic strictures or removal of intrahepatic stones as required (see Figs 4.4 and 4.5, pages 57 and 58). However, if liver transplantation is required, the operation may be technically more difficult due to the previous surgery in the area.

Both percutaneous and endoscopic dilation of dominant strictures has been performed. The timing of intervention is difficult and there are two schools of thought. One group waits until the patient has persistent jaundice and indications for liver transplantation. The other approach is to intervene earlier before the patient develops chronic liver disease secondary to prolonged obstruction. It is hoped that the benefits of treatment will outweigh any complications of the procedure. As sclerosing cholangitis is uncommon and has a variable clinical course, it has not been possible to compare these two clinical strategies adequately.

Huibregtse (1988) attempted stenting in 17 patients with sclerosing cholangitis and was successful in all but one. Stents were replaced electively every 3 months and the strictures reviewed cholangiographically after 1 year. Improvement was seen in 13 patients and after 12 months the stent could be removed permanently in eight patients.

Johnson *et al.* (1987) managed a group of 10 patients with a combination of endoscopic sphincterotomy (10), balloon dilation of dominant strictures (5) and dilation combined with stenting (3). Over a follow-up period of 19 months liver function tests improved and the frequency of cholangitis decreased.

Lombard *et al.* (1991) treated six deeply jaundiced patients (mean bilirubin 266 μmol/l). Dominant strictures were dilated to 8 mm diameter and a 10 Fr stent was inserted. Jaundice improved in all patients and five showed sustained biochemical improvement. Stents were removed without replacement in four patients 6 to 13 months after initial insertion. Cholangiography showed impressive improvements in stricture diameter.

Figure 6.7a shows a typical cholangiogram of sclerosing cholangitis with a dominant stricture of the common bile duct together with strictures of the left and right hepatic ducts. The patient presented with a persistently elevated bilirubin (200 μg/l) with associated malaise and weight loss. The common duct stricture was dilated and a stent inserted (Fig. 6.7b). His jaundice resolved and weight increased. Twelve months later a repeat cholangiogram showed the stricture was much improved (Fig. 6.7c) and the stent was removed. His bilirubin remains normal 12 months later.

Postsurgical strictures

These usually present with jaundice, cholangitis or in the immediate postoperative period as a biliary fistula. Patients with symptoms should be treated promptly. Long-standing biliary strictures may lead to secondary biliary cirrhosis with associated portal hypertension. Surgical repair by an experienced hepatobiliary surgeon gives excellent results and is the treatment of choice in young fit patients. Stenting should be considered in patients who are high risks for surgery and is probably the treatment of choice in the small subgroup who present with a biliary leak or persistent biliary drainage immediately postoperatively. Early repeat surgery is often hazardous and a biliary stent will quickly restore internal biliary drainage and the fistula will dry up in 1 to 2 days. The situation can then be reassessed when the patient has recovered from the initial surgery.

The management of strictures in other areas such as the urethra and oesophagus suggests that recurrence could be a major problem after a single dilation. The optimal endoscopic treatment has not yet been established, but the best results have been obtained with a combination of dilation followed by stenting. The stricture is dilated initially and one or more stents are inserted. These are replaced every 3 months and the stricture is reassessed after 12 months. Often the stents can be removed permanently at this time and the stricture remains dilated. These strictures may be technically difficult to dilate

Table 6.4 *Results of endoscopic management of postoperative benign biliary strictures.*

	Davids et al., 1992 n = 70	Geenan et al., 1989 n = 25
Age (years)	58	47
Stricture Location (%)		
Hilar	60	
Mid duct	20	68
Low	20	32
Successful Dilatation (%)	94	88
Early Complications (%)	9	8
Follow-up (years)	3½	4
Late Recurrences (%)	17	?

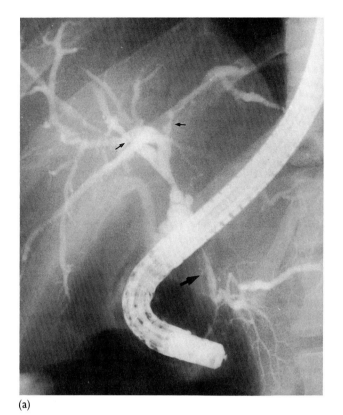

(a)

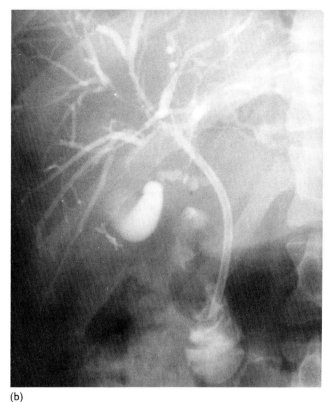

(b)

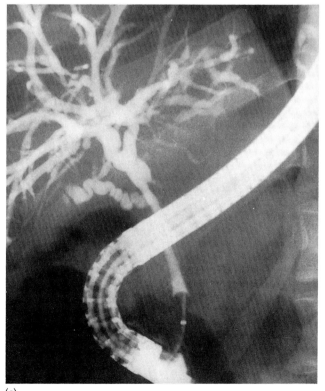

(c)

Fig. 6.7

(a) Endoscopic retrograde cholangiogram showing a long tight stricture of the common bile duct and common hepatic duct. Multiple diverticula are seen at the lower end (large arrow). There are other strictures at the origin of the left hepatic duct, and in the right intrahepatic ducts (two small arrows). These features are typical of sclerosing cholangitis. (b) A 10 Fr stent has been placed across the dominant stricture shown in Fig. 6.7a. (c) The stent has been removed and a balloon catheter placed in the lower end of the bile duct to occlude the sphincterotomy. The cholangiogram shows the stricture has improved markedly and the common bile duct now measures 3 mm diameter.

slightly. Davids *et al.* (1992) attempted to place at least two 10 Fr stents side by side through the stricture to provide maximal dilation. Geenan *et al.* (1989) used a single 10 Fr stent and dilated the stricture up to 10 mm in diameter several times during the follow-up period. These results are encouraging, but the follow-up period is relatively short. Surgical experience has shown that strictures may recur even 10–15 years postprocedure.

CONCLUSIONS

Endoscopic stenting has revolutionized the palliation of biliary obstruction due to malignancy and is now being used more frequently in benign stric-

and often require two or more attempts to complete the procedure.

The results of two large endoscopic series are summarized in Table 6.4. The techniques used differ tures. Equipment and techniques are continually being refined and results improved. Similar exciting advances are being made in the fields of percutaneous transhepatic intervention and hepatobiliary surgery. Patients with biliary obstruction are now best managed by a team comprising individuals with an active interest and expertise in hepatobiliary surgery, endoscopic and percutaneous transhepatic intervention.

REFERENCES

DAVIDS PHP, RAUWS EAJ, COENE PPLO, TYTGAT GNJ, HUIBREGTSE K (1992). Endoscopic stenting for post operative biliary strictures. *Gastrointestinal Endoscopy* **38**: 12–18.

DEVIERE J, BAIZE M, DE TOEUF J, CREMER M (1988). Long term follow-up of patients with hilar malignant stricture treated by endoscopic internal biliary drainage. *Gastrointestinal Endoscopy* **34**: 95–101.

DOOLEY JS, HAMILTON-MILLER JMT, BRUMFITT W, SHERLOCK S (1984). Antibiotics in the treatment of biliary infection. *Gut* **25**: 988–98.

FRAUMENI JF JR (1975). Cancers of the pancreas and biliary tract. Epidemiological considerations. *Cancer Research* **35**: 3437–46.

GEENAN DJ, GEENAN JE, HOGAN WJ, SCHENCK J, VENU RP, JOHNSON GK, JACKSON A (1989). Endoscopic therapy for benign bile duct strictures. *Gastrointestinal Endoscopy* **35**: 367–71.

HUIBREGTSE K (1988). *Endoscopic Biliary and Pancreatic Drainage*, pp. 104–9. George-Theime Verlag, Stuttgart, New York.

HUIBREGTSE K, KATON RM, COENE PP, TYTGAT GNJ (1987). Endoscopic palliative treatment in pancreatic cancer. *Gastrointestinal Endoscopy* **32**: 334–8.

JOHNSON GK, GEENAN JE, VENU RP, HOGAN WJ (1987). Endoscopic treatment of biliary duct strictures in sclerosing cholangitis: follow-up assessment of a new therapeutic approach. *Gastrointestinal Endoscopy* **33**: 9–12.

LOMBARD M, FARRANT M, KARANI J, WESTABY D, WILLIAMS R (1991). Improving biliary-enteric drainage in primary sclerosing cholangitis: experience with endoscopic methods. *Gut* **32**: 1364–5.

SHEPHERD HA, ROYLE G, ROSS APR, DIBA A, ARTHUR M, COLIN-JONES D (1988). Endoscopic biliary endoprosthesis in the palliation of malignant obstruction of the distal bile duct; a randomized trial. *British Journal of Surgery* **75**: 1166–8.

SMITH AC, DOWSETT JF, RUSSELL RCG, HATFIELD AR, WILLIAMS S, MACRAE KD, HOUGHTON J, LENNON CA (1990). Endoscopic stenting versus bypass surgery for primary low malignant biliary obstruction. A prospective randomized trial in 204 patients. Published in the *Proceedings of World Congress of Gastroenterology*.

SPEER AG, COTTON PB, RUSSELL RCG, MASON RM, HATFIELD AR, LEUNG JWC, MACRAE KD, HOUGHTON J, LENNON CA (1987). Randomized trial of endoscopic versus percutaneous stent insertion in malignant obstructive jaundice. *Lancet* **ii**: 57–62.

SPEER AG, COTTON PB (1988). Endoscopic treatment of pancreatic cancer. *International Journal of Pancreatology* **3**: S147–58.

SPEER AG, COTTON PB (1989). Endoscopic stent placement in malignant biliary obstruction. In: *ERCP: Diagnostic and Therapeutic Application*. Edited by Jacobson IM. Elsevier Science Publishing Co., New York.

2

BIOPSY AND ABSCESS DRAINAGE

Computed tomographic-guided and ultrasound-guided intra-abdominal biopsy and abscess drainage

Eugene Y. Yeung

Introduction   100

Image-guided biopsy   100

Radiological guidance techniques   104

Image-guided drainage   108

References   116

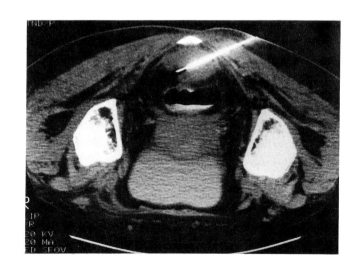

INTRODUCTION

The radiologically guided percutaneous needle biopsy is of proven value in the investigation of intra-abdominal pathology. Percutaneous drainage has replaced surgery as the treatment of choice for most intra-abdominal abscesses and collections. The rise in popularity of these techniques can be traced to the tremendous advances in ultrasound and computed tomographic technology, needle and catheter design, and understanding of percutaneous techniques. This chapter describes percutaneous techniques with special emphasis on the elements that ensure their success.

IMAGE-GUIDED BIOPSY

Indications and contraindications

Almost any radiologically detected abdominal abnormality is suitable for image-guided biopsy. Percutaneous biopsy can provide information about the presence, nature and extent of disease. Biopsy is performed most often to investigate malignancy, although it is also of value in detecting benign or infectious disease.

Percutaneous biopsy is contraindicated in patients who have an increased risk of haemorrhage, sepsis or tissue damage (Table 7.1). However, most contra-indications are relative, and the risk versus benefit to the patient must be considered.

Patient preparation

An intravenous (i.v.) line should be inserted and i.v. analgesia given before the procedure. Patients with abnormal haemostasis (i.e. prothrombin time ratio >1.5 and platelet count <50 000/mm^3) are given the appropriate prophylaxis (vitamin K) and blood product replacement. Patients who have severe, uncorrectable coagulopathies and require percutaneous liver biopsy should be considered for a plugged (Riley *et al.*, 1984; Allison and Adam, 1988) or transjugular (Gamble *et al.*, 1985) biopsy (see Chapter 8).

Antibiotic prophylaxis should be considered in immunocompromised patients or if there is a high probability of traversing a contaminated field (e.g. transcolonic biopsy).

Table 7.1 *Relative Contraindications to Percutaneous Biopsy*

- Abnormal haemostasis (prothrombin time ratio >1.5; bleeding time >10 minutes; platelet count <50 000/mm^3)
- Highly vascular lesions (e.g., liver haemangioma and hepatoma)
- Immunosuppression (especially if a contaminated field is traversed, e.g., colon)
- Uncooperative patient
- Gross ascites

If there is clinical suspicion that the lesion is vascular (e.g. liver haemangioma), a contrast-enhanced computed tomographic (CT) scan should be performed before the biopsy. If high vascularity is confirmed, the need for biopsy should be reassessed. Biopsy of possible hormone-secreting (e.g. carcinoid) tumours should be performed only after appropriate treatment with blocking agents (Bissonnette *et al.*, 1990).

Choice of needle

Biopsy needles can be divided into three groups (Figs 7.1 and 7.2): small-gauge aspiration needles (e.g. Chiba), small-gauge core-biopsy needles [Turner, Franseen (Cook, Bloomington, IN)], and larger core-biopsy needles, for example, 18 G Biopty

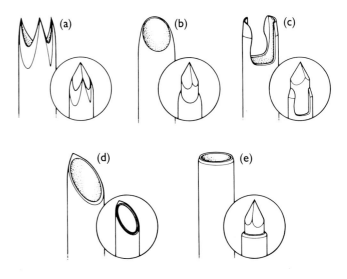

Fig. 7.1 *Small-gauge biopsy needles*
(a) Franseen (core). (b) Turner (core). (c) EZ-EM (core). (d) Chiba (aspiration). (e) Standard trocar (aspiration). Insets show needles with inner stylets in place.

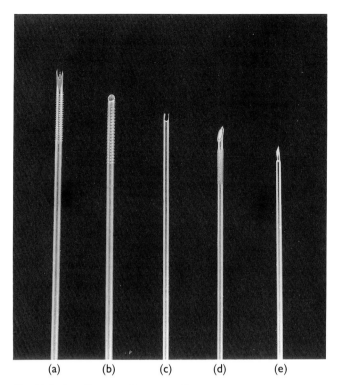

Fig. 7.2 *Small-gauge biopsy needles*
From left to right: (a) Franseen (b) Turner (c) EZ-EM (d) Standard Chiba (e) Standard Trocar. Needles (a–c) have stylets removed to show cutting surfaces. Needles (a), (b) and (d) have a ridged shaft for enhanced ultrasound visualization.

(Bard, Coppington, GA), 14G Trucut (Baxter Health Care Valencia, CA). The following factors should be considered when choosing the needle.

Aspiration needles obtain only cytological samples, which are often scant. Therefore, multiple biopsy passes are usually required and an experienced cytopathologist is needed for accurate interpretation. Since histological samples are not obtained, aspiration biopsy is not usually suitable for diagnosing benign lesions and certain malignancies (e.g. lymphoma).

Small-gauge core-biopsy needles are designed to provide small cores of tissue that allow histological analysis. Combined cytological and histological analysis increases the overall diagnostic accuracy (Wittenberg *et al.*, 1982). Large core-biopsy needles provide good histological samples although the risk of complications may be higher.

Increasing the calibre of the needle increases the sample size (Andriole *et al.*, 1983; Haaga *et al.*, 1983). However, no clear cut advantage of one needle tip configuration over another has been shown.

There is increasing interest in the use of a spring-propelled 18 G cutting needle [Biopty (Lindgren, 1982)] for biopsy. This works like the Trucut needle but has a smaller calibre and a rapid action. This device provides excellent histological samples with few complications (Parker *et al.*, 1989).

Choosing the biopsy route

Generally, the shortest route from the skin to the lesion is chosen, providing no vital structures (e.g. the spleen or lung) are interposed. An exception is the biopsy of peripheral vascular hepatic lesions. There are scattered reports of life-threatening haemorrhage from biopsy of liver haemangioma and hepatomas. A contrast CT scan identifies such lesions by their heightened enhancement relative to surrounding liver. If biopsy is contemplated in these lesions, the biopsy needle path should be carefully chosen to traverse normal liver between the liver capsule and target (Fig. 7.3). This will tamponade any haemorrhage (Solbiati *et al.*, 1985).

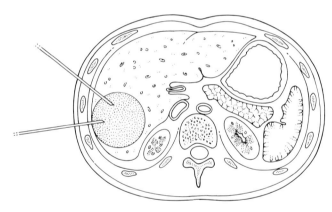

Fig. 7.3
Biopsy of peripheral liver lesions. The biopsy needle should traverse normal liver before entering the lesion (i.e. route A is preferred over route B).

Directing the needle

Although image guidance is used, the needle trajectory usually requires some readjustment as it is advanced towards the target. A needle with a bevelled tip deflects tissue asymmetrically as it is advanced, tending to deviate the needle from its path. Thus, the needle can be redirected by partially withdrawing it, reorienting its tip and readvancing it (Fig. 7.4) (Horton *et al.*, 1980).

If multiple biopsy passes are required, different parts of the lesion should be sampled using the coaxial or tandem method. In the coaxial method, a fine-bore needle is inserted into the lesion and a

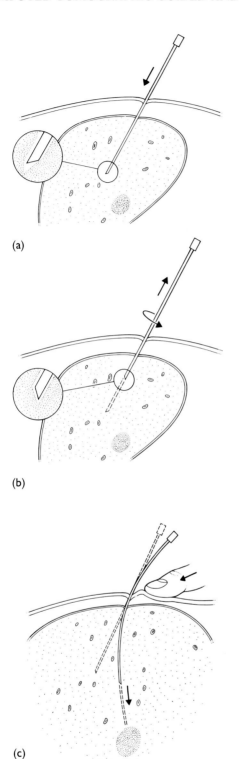

(a)

(b)

(c)

Fig. 7.4 *Technique of redirecting a bevelled needle during biopsy of a liver lesion*
(a) The needle trajectory needs readjustment (see inset). (b) The needle is withdrawn to the periphery of the liver and rotated to reorient the bevel (see inset). (c) The needle is readvanced while the shaft of the needle at the skin entry point is deviated in the direction shown.

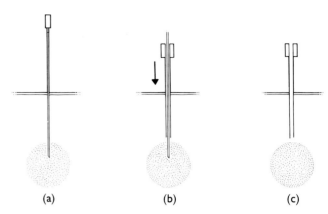

(a) (b) (c)

Fig. 7.5 *Coaxial method of needle placement*
(a) 21/22 G thin needle is inserted into the lesion. (b) The 18 G needle is inserted over the thin needle to the periphery of the lesion. (c) After the biopsy is taken, the thin needle is withdrawn leaving the 18 G needle in place. Further biopsies are taken through the 18 G needle.

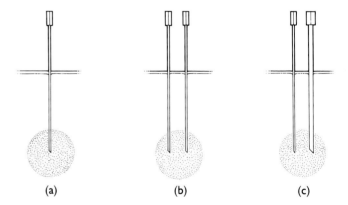

(a) (b) (c)

Fig. 7.6 *Tandem method of needle placement*
(a) The initial biopsy needle is inserted into the lesion. (b) Using the first needle as a reference, a second, similar-sized needle is inserted 0.5–1.0 cm away from the first needle. Up to four needles can be inserted in this way. The entry sites of the second and subsequent needles should not be too close because their shafts will interfere with each other causing misdirection. (c) The second needle may be of larger calibre (e.g. Trucut needle) for histological sampling. This technique is also useful for placing single stab trocar drainage catheters.

larger outer needle is inserted over the fine needle to the edge of the lesion (Fig. 7.5). The sample is taken using the fine needle. Further biopsies are performed through the outer needle, which acts as a guide. This method requires special needles that have either a preloaded outer needle or a removable hub on the inner needle (McGahan, 1984).

In the tandem method, the needle inserted initially is left in position even if it is not in the lesion (Fig. 7.6). Its direction and depth is then used as a guide to insert one to three additional needles (Fig. 7.7) (Ferrucci *et al.*, 1980).

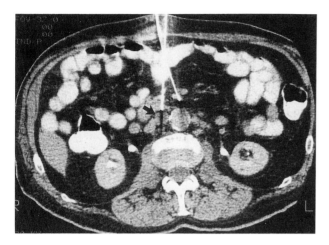

Fig. 7.7
Tandem method of para-aortic node biopsy. Computed tomographic scan showing the first needle was deflected from its chosen trajectory and lies anterior to the aorta. Using the first needle as a guide, a second needle was placed accurately for subsequent biopsy of the para-aortic node (arrow). Note the shadow artefact of the second needle which confirms the scan is through the needle tip. Note also the vertical and axial needle trajectory.

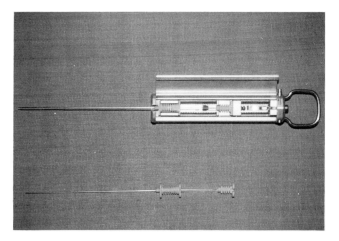

Fig. 7.8
18G Biopty gun. The gun mechanism and attached biopsy needle are shown. The needle with the plastic sheath interposed between its hubs is shown below. This is inserted as one unit and, after correct placement has been confirmed, the gun mechanism is attached for biopsy.

Sampling technique

For aspiration biopsies, a 20 ml syringe is attached and at least 10 ml of suction applied (Heuftle and Haaga, 1986) while an up-and-down and twisting motion with 1–2 cm of excursion is used. Suction is then released and the needle withdrawn. Recently, Fagelman and Chess (1990) suggested that removing the inner stylet and oscillating the needle without attaching a syringe provides a similar diagnostic yield with less chance of contamination with blood.

When the Biopty gun is used, the needle tip should be positioned at the near edge of the target lesion since the needle thrusts forwards to take the biopsy (Parker *et al.*, 1989). Various needle travels are available (e.g. 1.5 or 2 cm).

If the Biopty gun is used with CT guidance, CT gantry clearance restrictions require the needle to be detached from the gun mechanism during insertion into the target and the gun to be reattached when the needle is in position. Because there is no locking mechanism between the two components of the Biopty needle (trocar and cannula), an appropriate length of the sterile plastic sheath that accompanies the needle is interposed between their hubs so they may be inserted as one unit. A slit can be made along the length of the sheath so it can be removed without withdrawing the inner trocar (Fig. 7.8).

Aftercare

Minimal care is required after percutaneous biopsy. Patients should be on bed rest for at least 2 hours but, as long as vital signs remain stable, they may then gradually resume normal activities.

Results and complications

Overall accuracy rates of percutaneous abdominal biopsy exceed 80% in most series (Ferrucci *et al.*, 1980; Welch *et al.*, 1989). False negative biopsies may be due to an off-target biopsy or sampling of the necrotic or fibrous portion of a tumour (e.g. pancreatic carcinoma). Thus, a negative biopsy should be repeated when there is strong clinical suspicion of malignancy. The use of a larger gauge or core-biopsy needle should also improve the yield.

Complication rates from percutaneous abdominal biopsy are extremely low, especially when fine-gauge needles are used, and most series report rates of less than 1.5% (Lees *et al.*, 1985; Welch *et al.*, 1989).

RADIOLOGICAL GUIDANCE TECHNIQUES

Ultrasound guidance

Ultrasound has become the preferred imaging modality for guided biopsy because of its ready availability and relatively low machine cost. Although ultrasound guidance may be performed using an ultrasound transducer with a biopsy guide attachment, this technique is cumbersome and restrictive, and the freehand real-time method is usually preferred.

In the freehand technique, the biopsy needle is imaged in real-time as it advances from the skin entry point to the target (Matalon and Silver, 1990; Yeung *et al.*, 1990). Any deviation from the required path is readily detected. The technique requires the simultaneous use of both hands, one to hold the ultrasound transducer and the other to guide the biopsy needle. Good hand and eye coordination and ambidexterity are required.

TECHNIQUE

The optimal skin entry point and proposed needle path to the target lesion are first chosen by scanning the patient. The absence of interposed vital structures (e.g. spleen or lung) in the path of the needle is confirmed by scanning the target lesion from the proposed skin entry point. The position of the transducer during biopsy is then chosen (Fig. 7.9). This should allow simultaneous imaging of the target lesion and the advancing needle tip and shaft. Thus, the ultrasound scan plane should be aligned with the plane of the needle path. To aid accurate placement of the needle in the target further, the transducer position should be adjusted so that the target is in the centre of the image obtained. The vertical axis of the transducer points directly to the target and provides an important visual aid for precise needle placement. Local anaesthetic is given and the needle is inserted and advanced towards the target during imaging. Minor adjustments to the needle trajectory are made when necessary.

CHOICE OF TRANSDUCER

Sector transducers are physically more compact than linear arrays and are preferred in regions with limited access (e.g. intercostal spaces) or if an angled approach is required. Near field imaging is, however, inferior to that of a linear array transducer,

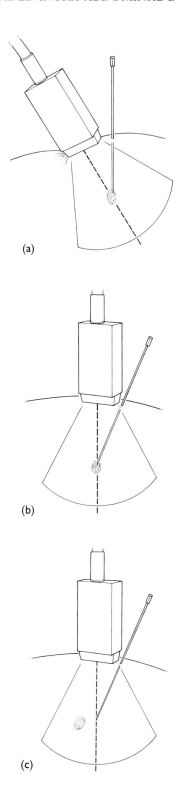

Fig. 7.9
Technique for ultrasound guidance. The needle path, target and scan plane should all be aligned to allow imaging of the advancing needle. (a) The needle trajectory is perpendicular to the skin with the transducer angled. (b) The transducer is perpendicular to the skin with the needle angled. (c) When the target is not in the centre of the ultrasound image, the vertical axis of the transducer does not point directly at it.

which is more suitable for superficial targets. The physical shape of the sector transducer is also important. A rectangular transducer is preferred to a round one because the scanning plane is more easily identified.

TRANSDUCER STERILITY

The ultrasound transducer should be covered by a sterile drape during biopsy. Plastic or rubber bags filled with scanning gel have been used to maintain sterility. Alternatively, a sheet of Opsite dressing (Smith and Nephew, Hull, UK) 30 cm × 28 cm may be used to dress the transducer (Fig. 7.10). The orientation of the transducer head is clearly visible through the Opsite dressing, allowing prompt identification of the scan plane.

VISUALIZATION OF THE NEEDLE

Poor visualization of the needle tip is often cited as a problem. Most needles, however, are sufficiently visible sonographically as long as the scan plane is correct. The most common cause of nonvisualization of the tip is oblique sectioning (i.e. the target lesion and needle are not in the same plane). Such misalignment is obvious from the orientation of the exposed part of the needle shaft relative to the axis of the ultrasound transducer. Realignment can be achieved by adjusting the transducer position so the

target is in the centre of the image and the scan plane bisects the needle entry point. The appropriate correcting manoeuvre can then be gauged. When searching for the needle tip, avoid haphazard transducer movements. Orderly transducer movements give a better indication of the relationship of the needle tip to the target lesion (Fig. 7.11).

Techniques for enhancing visualization of the needle include bobbing the needle or moving the stylet while scanning. Extrareflective needles specifically designed for ultrasound guidance are commercially available if required (Fig. 7.2). Injection of a small amount of air through the needle has been advocated to identify its tip (Lee and Knochel, 1982). Unfortunately, this often degrades the ultrasound image and reduces visibility of the target so that subsequent attempts to visualize or identify the needle tip are more difficult.

Computed tomographic guidance

TECHNIQUE

The patient is placed in the position for biopsy (e.g. prone, supine or oblique) and preliminary scans through the target area taken. Respiration should be suspended during scan acquisition, preferably at the resting expiratory level, since patients find this more comfortable. Intravenous contrast medium should be given to enhance visualization of the lesion, assess its vascularity and highlight surrounding structures. An appropriate needle path and skin entry point are then chosen from the acquired images. A vertical or horizontal trajectory in the axial plane is preferred because it obviates 'triangulation' (see below) and allows the entire needle to be visualized on a single CT image (Fig. 7.6). The skin entry point is identified on the patient by taping a radio-opaque marker strip on the skin and rescanning the area (Fig. 7.12). A simple radio-opaque marker strip can be made by placing metal rods (e.g. metal paper clips cut into 2 cm lengths) at 1 cm intervals and sealed on a piece of tape. A local anaesthetic is then given and the needle is inserted half the distance to the target. A repeat scan at this time shows any deviation from the required trajectory, allowing readjustment to needle positioning.

Frequently, angled approaches (e.g. cephalocaudal) may be required because vital structures surround the target. On some CT scanners, the gantry may be tilted to allow direct visualization of the needle pathway (Yueh *et al.*, 1989). If this is not feasible, several methods that rely on basic geometric principles ranging from simple visual

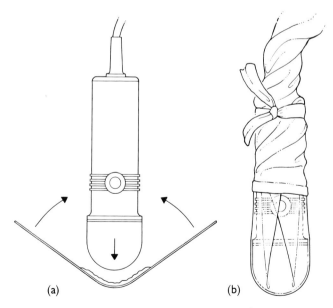

(a) (b)

Fig. 7.10 *Technique to maintain transducer sterility* (a) The operator opens the Opsite dressing. An assistant applies scanning gel to the centre of the dressing and then places the transducer vertically onto it. (b) The operator wraps the Opsite around the transducer and covers the cable with a sterile towel secured by a gauze dressing.

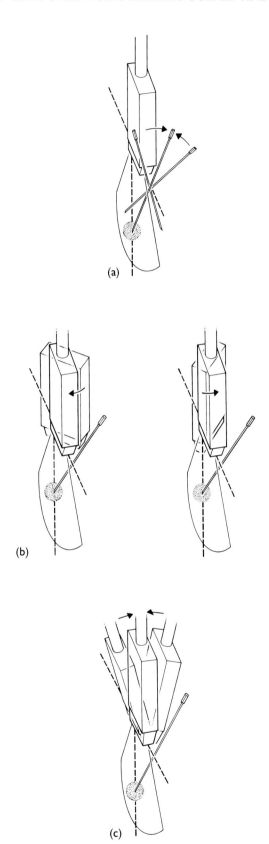

Fig. 7.11 *Nonvisualization of the needle due to oblique sectioning*
(a) Misalignment of the needle with correct transducer position.
(b,c) Misalignment of the transducer position.

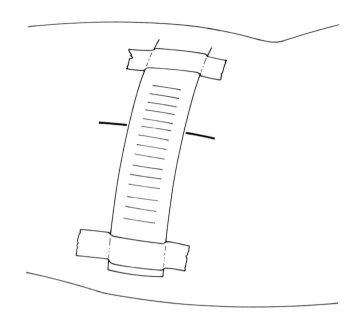

Fig. 7.12
A home-made radiopaque marker strip placed on the patient while a computed tomographic scan is done.

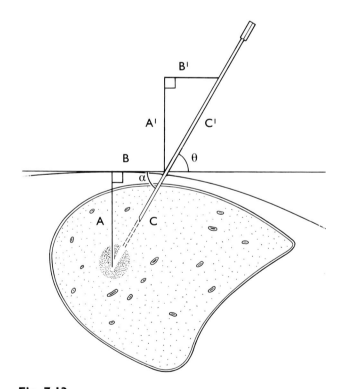

Fig. 7.13
Methods for angled computed tomographic (CT) biopsy.
Right-angled triangle ABC is congruent to triangle A'B'C'. A and B are measured on the CT monitor image using electronic callipers. In the visual method, A' is reproduced on the patient by holding a ruler vertically at the patient's skin. B' is estimated or a second ruler used. In the calculated method, angle θ equals angle α equals tangent A/B. Angle θ is reproduced on the patient using a protractor or variable angle measurer (goniometer).

techniques to calculation of actual angles of approach may be used (Fig. 7.13).

Visual method

The visual method (Axel, 1984) relies on the premise that the right-angled triangle formed by the shaft of the needle positioned in the target, an imaginary vertical line from the target to the skin surface and a horizontal line from that vertical line to the skin entry point is exactly congruent to the triangle formed by the shaft of the needle outside the body, an imaginary line projected through the skin entry point and the horizontal line from the shaft of the needle to the vertical line. To estimate the angle of approach, measured distances A and B are reproduced by the external triangle by holding a ruler vertically over the skin entry point. At height A^1, the needle shaft should be B^1 distance from the vertical line.

Calculated method

Based on the same triangle, the angle of approach is calculated by the formula (van Sonnenberg *et al.*, 1981):

Angle θ equals angle α, which equals tangent A/B

The needle is inserted along the calculated angle using a goniometer (variable angle measurer) or a simple protractor placed at the skin entry point for reference.

Fluoroscopic guidance

When the target is made visible with contrast medium (e.g. abnormal lymph node seen on lymphangiography or bile duct stricture with an indwelling biliary stent), fluoroscopy is suitable for biopsy guidance (Fig. 7.14). For successful fluoroscopically guided biopsy, the skin entry site should be directly over the target lesion in line with the X-ray beam so that the needle is imaged down its vertical axis. Deviations from this trajectory are easily detected. The needle tip depth in relation to the target may be assessed by either angling the X-ray beam or placing the patient at an oblique angle (Fig. 7.15).

Choice of imaging modality

Computed tomography or ultrasound may be used with equal success in most instances. The decision often depends on the availability of the imaging modality and the preference of the operator. Each technique is, however, more suited to certain situa-

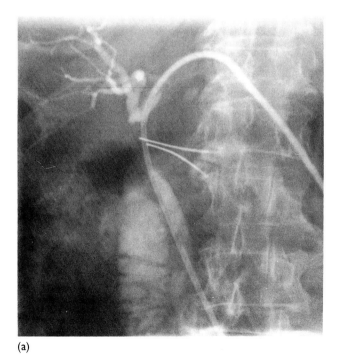

(a)

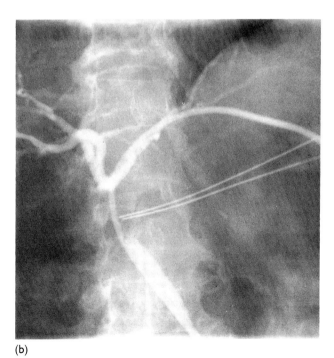

(b)

Fig. 7.14 *Fluoroscopically guided biopsy of a liver hilar mass*
(a) Two fine-gauge needles inserted into the presumed site of the mass are made 'visible' by contrast medium injection through the percutaneous biliary drainage tube. (b) Putting the patient at an oblique angle to the left confirms that both needles are well positioned slightly anterior to the drainage tube.

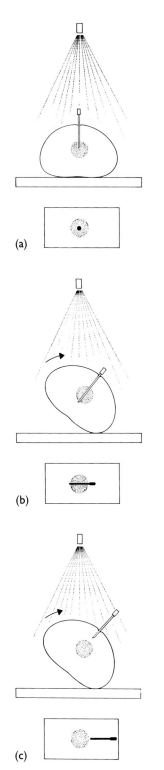

Fig. 7.15 *Assessing the depth of the biopsy needle by fluoroscopy*
(a) The needle is imaged along its vertical axis to the target. (b) If the needle is too deep, placing the patient at an oblique angle deviates the needle tip so that it appears to cross the target on the fluoroscopic image. (c) If the needle is too superficial, placing the patient at an oblique angle separates the lesion and needle tip. To eliminate parallax, the fluoroscopic image should be adjusted so that the needle and target are in the centre.

tions (Fig. 7.16). Small targets affected by respiratory motion are often biopsied more readily using ultrasound. Similarly, because of the multiple scan planes possible with ultrasound, angled approaches (e.g. craniocaudal angulation) are easier. On the other hand, ultrasound biopsy may be safe only if the target lesion is well visualized. Thus, deep structures that may be obscured by bowel gas (e.g. pancreatic tail lesions) should be biopsied under CT guidance (Table 7.2).

IMAGE-GUIDED DRAINAGE

Indications and contraindications

Almost any intra-abdominal fluid collection demonstrated by either CT or ultrasound may be considered for drainage.

The standard indication for percutaneous drainage is a unilocular or discrete fluid collection. However, because of its low morbidity, the indications have been extended to include collections of various kinds (e.g. necrotic tumour, haematomas and phlegmonous masses). Although percutaneous drainage may not totally eradicate these complex collections, the procedure may be attempted as a palliative procedure, to gain time and allow improvement in the patient's general state of health so that definitive surgical drainage and debridement can be performed.

There are few contraindications to percutaneous drainage. Most notable are abnormal haemostasis and an unsafe access route (e.g. bowel or lung).

Patient preparation

A coagulation profile is obtained. Intravenous access is mandatory and i.v. analgesia should be given. Patients not already receiving antibiotics should be given broad-spectrum coverage before and after drainage.

Radiological guidance techniques

The image guidance techniques are as described above. Percutaneous drainage of intra-abdominal collections is most often performed using ultrasound or CT guidance either alone or with fluoroscopy. Fluoroscopic guidance may be used on its own when

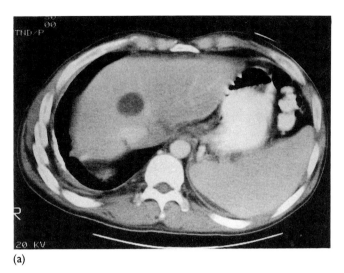

(a)

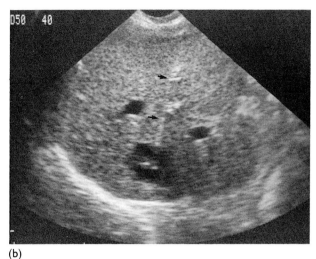

(b)

Fig. 7.16 *Choosing between ultrasound and computed tomographic (CT) guidance*
(a) Computed tomography of the liver shows a cystic lesion in the superior part of the right lobe. Lung surrounds the liver making CT guidance difficult, especially if respiratory motion is marked. (b) Ultrasound guidance simplifies the biopsy. In this patient, the approach was from the right subcostal region. The target is in the centre of the image and the scan plane is in line with the needle shaft (arrows).

Table 7.2 *Comparison of the Relative Advantages and Disadvantages of CT- and Ultrasound-guidance*

	CT-guidance	**Ultrasound-guidance**
Advantages	Images are easy to comprehend (3-dimensional)	Low cost
		Often extremely fast
	Excellent visualization of deep structures	Real time (respiratory motion not a problem)
	Images are not obscured by bowel gas or bone	Angled approaches are usually simple to perform
		Vascular structures are readily identified
		No x-rays are used
Disadvantages	May be time consuming	Competence is more difficult to achieve
	Craniocaudal approaches may be difficult	Poorly visualized targets cannot safely be biopsied
	CT time is expensive	
	X-rays are used	

the collection is superficial and large and its position has been located by a prior ultrasound or CT examination.

Ultrasound guidance is usually preferred over CT because of its real-time properties and because it can be readily combined with fluoroscopic guidance. Ultrasound is particularly useful in the drainage of superficial collections and those requiring an angled approach (e.g. subphrenic, high hepatic and splenic abscesses).

Deep abdominal and pelvic collections are poorly visualized by ultrasound and CT guidance is preferred. However, combined ultrasound and fluoroscopic guidance may still be used to drain poorly visible collections as long as a CT scan is available to help plan a safe access route. Computed tomographic-guided drainage is not real time and the guidewire, track dilators and drainage catheter are inserted by feel alone. Therefore, extra care is required to avoid misplacement of the catheter. Repeated CT slices may be performed to monitor each stage of the procedure.

Fluoroscopic guidance may also be used with CT by positioning a mobile image intensifier between the CT table and the gantry.

Choosing the drainage route

The shortest route from the skin entry point to the collection should be chosen as long as no vital structures are transgressed. Inadvertent transgression of vital structures (e.g. bowel, vessels and pleural space) by the drainage catheter may have dire consequences. A retroperitoneal as opposed to an intraperitoneal approach is preferable to reduce the risk of peritoneal spillage of infected material.

Whenever possible, the most dependent portion of the collection should be entered.

Diagnostic needle aspiration

The CT and ultrasound appearances of intra-abdominal fluid collections cannot often reliably differentiate between abscess, sterile collections and organizing haematomas. When the decision to drain a collection depends on its nature or whether it is infected, needle aspiration may be performed before drainage. If the aspirated fluid is not overtly purulent, it should be sent for immediate Gram staining. The sample must be taken from the dependent portion of the collection: abscesses have a tendency to layer so that sampling the relatively clear supernatant component may result in a false negative result. If fluid cannot be aspirated with a small-gauged needle, progressively larger needles or sheaths up to 18 G should be used.

Biochemical analysis may also help characterize the fluid (e.g. identification of a bile leak after a cholecystectomy or injury to the pancreas after splenectomy by determining the amylase level).

Drainage technique

Drainage catheters may be inserted using the Seldinger guidewire exchange technique or via a single-step trocar drainage catheter system.

For the Seldinger technique, the choice of introduction systems depends mainly on personal preference, for example, coaxial systems using 0.018 inch wires like the Cope and Hawkins introducer set (Hawkins *et al.*, 1988), Mitty-Pollack 22/18 G co-axial needle (Cook, Bloomington, IN) and an 18 G sheathed Lunderquist needle. A tandem approach to needle insertion may also be used (Fig. 7.6). Confirmation of the needle tip's position in the collection is made by injecting a small volume of contrast medium; large injections to delineate the cavity should be avoided at this stage. Once a 0.038 inch guidewire has been inserted, the track is dilated and the drainage catheter inserted. If tissue resistance is encountered, an appropriately sized peel-away sheath (Cook) may be introduced first and the drainage catheter inserted through this. Most commercially available drainage catheters come supplied with an inner metallic stiffener to aid insertion.

The drainage catheter and side holes should be advanced or coiled along the entire length of the abscess cavity to allow complete evacuation. If necessary, extra side holes should be cut. Care should be taken to ensure that all the side holes are within the abscess to prevent spillage.

One-step trocar catheter systems are used when collections are large and superficial and there is little risk of misdirection. When using a trocar catheter [e.g. van Sonnenberg Sump (Meditech, Watertown, MA)] a deeper and more generous skin incision should be made to facilitate entry.

Multiple adjacent abscess cavities may require several drainage catheters for adequate treatment. Septated or loculated abscess cavities may also cause incomplete drainage. Loculations may be disrupted by the use of a guidewire or a catheter.

Catheter selection

The choice of catheter depends mainly on the viscosity of the fluid. Essentially, the catheter should allow steady drainage. Thus, large bore catheters (12–18 Fr) with large side holes should be used if the material aspirated is thick or particulate. Catheters of 8–10 Fr are adequate for nonviscous fluids such as bilomas and loculated ascites. Catheters less than 8 Fr are not effective (Gobien *et al.*, 1985). Double-lumen sump drains [e.g. van Sonnenberg (Meditech), Ring MacLean (Cook)] are favoured by some (van Sonnenberg *et al.*, 1982). These have a large main lumen that allows steady drainage of fluid while the second (sump) smaller lumen allows room air to percolate into the cavity, theoretically preventing adherence of the abscess wall to the catheter side holes. Irrigation fluid (saline or proteolytic agents) may also be infused into the cavity through the second lumen. In practice, however, these catheters probably give no real advantage over single-lumen catheters of similar size. Nasogastric tubes (e.g. Argyle, St Louis, MO) with custom-cut side holes are probably as effective and are available at a fraction of the cost.

Evacuation of cavity contents

Insertion of the drainage catheter should be followed by as complete evacuation of the cavity contents as possible (Mueller *et al.*, 1984). Altering the patient's position and gently manipulating the catheter aids complete evacuation. Evacuation should be terminated, however, if the patient experiences pain or the aspirate becomes bloody. Sudden decompression of large collections may be extremely uncomfortable for the patient and may precipitate haemorrhage from the abscess wall. The abscess cavity may also be irrigated when abscess contents can no longer be aspirated. Irrigation should be carried out by gently injecting and then aspirating 20–50 ml aliquots of saline until the aspirate clears. Overzealous irrigation should be avoided, however, since it may precipitate bacteraemia. Proteolytic agents, antibiotics (van Sonnenberg *et al.*, 1982) and streptokinase (Vogelzang *et al.*, 1987) have been instilled into collections and haematomas to liquefy the contents, although data on their efficacy are limited.

Aftercare and catheter removal

Vital signs should be assessed every 30 min for at least 4 hours and then every hour for 4 more hours. The catheter may be irrigated at the bed side using 10–20 ml saline to maintain patency during postdrainage days. The patient's condition usually improves noticeably within 24–48 hours after successful drainage. Failure to improve usually means drainage is inadequate, the collection is loculated or another problem is the cause.

Continued drainage of 50 ml/day or more several days after the procedure often suggests a fistulous communication (e.g. with bowel, bile duct or pancreas), which can be confirmed by a contrast study. As long as the normal internal flow is unobstructed, these fistulae will close although drainage may continue for 4–6 weeks or more. The catheter can be removed once drainage has ceased and ultrasound or CT and a contrast study through the catheter confirms that the cavity has disappeared.

Results and complications

The overall success rate for percutaneous abscess drainage is reportedly above 77% (van Sonnenberg *et al.*, 1984; Lang *et al.*, 1986). Successful drainage of simple unilocular abscesses approaches 100%, but it is lower in patients with complex abscesses (e.g.

multiloculations, inflammatory masses and enteric or biliary fistulae).

Most large studies report complication rates between 0 and 15% overall (Gerzof *et al.*, 1981; van Sonnenberg *et al.*, 1984). The major complications include sepsis, haemorrhage and injury to adjacent structures. Minor complications include skin infection, catheter blockage and displacement. Careful planning of the approach to the collection should minimize the risk of inadvertent puncture of bowel, vascular structures and pleura. Even if bowel is traversed by passage of the catheter, it often responds to conservative management with few sequelae (Mueller *et al.*, 1985).

Pancreatic collections

The radiological management of pancreatic collections depends on the type (van Sonnenberg *et al.*, 1989).

PSEUDOCYSTS

Pseudocysts are walled-off pancreatic collections. They most commonly occur in the lesser sac region and usually form 4 to 6 weeks after acute pancreatitis. Although pseudocysts frequently resolve spontaneously, those larger than 5 cm in diameter often require intervention. An uninfected pseudocyst should be drained electively when the wall has become well-defined and acute pancreatitic inflammation has subsided (usually at least 4 weeks after the acute episode). Earlier drainage may be contemplated if the patient's symptoms (abdominal pain and upper gastrointestinal obstruction) are worsening or the pseudocyst is enlarging. Infected pseudocysts should be drained more urgently.

A transgastric approach (pseudocystogastrostomy) to drainage is preferred when feasible (Matzinger *et al.*, 1988). The procedure is like a percutaneous gastrostomy insertion (Figs 7.17 and 7.18). The stomach is distended with air through a nasogastric tube. The cyst is entered by puncture through the anterior abdominal wall and anterior and posterior walls of the stomach. Apposition of the anterior stomach wall to the abdominal wall is confirmed by lateral fluoroscopy. Needles such as the Mitty–Pollack or Cope introduction system may be used for puncture. The procedure is usually performed using fluoroscopic guidance although ultrasound or CT may be used.

The collection may be drained through a direct percutaneous approach if the transgastric route is not feasible. However, there is a small but significant

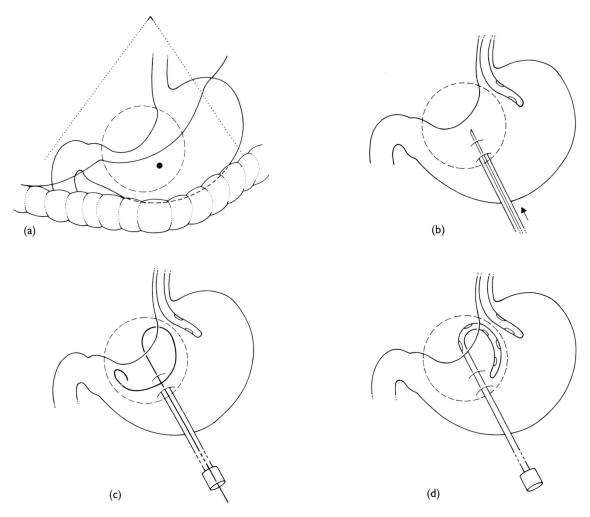

Fig. 7.17 *Schematic diagram of transgastric pseudocyst drainage*
(a) The pseudocyst usually lies behind the stomach in the lesser sac (dotted circle). The left lobe of the liver and transverse colon are also closely related. Computed tomography and ultrasound scans should always be done to delineate anatomy before the drainage procedure. The black dot indicates the preferred site for puncture. (b) Direct vertical stab of the stomach distended with air is performed using a 22/18 G Mitty-Pollack needle. The pseudocyst is entered by puncturing the posterior gastric wall. (c) Aspiration of cyst material or injection of contrast medium confirms correct needle placement. The 18 G outer needle is advanced into the cyst and a J-wire inserted. (d) A drainage catheter is inserted after appropriate dilation.

risk of development of a persistent cutaneous pancreatic fistula, which should not occur if the transgastric route is used.

The catheter is removed after CT and a catheter contrast medium injection confirms complete resolution of the collection with no residual loculated areas, and drainage volumes have decreased to below 10 ml/day. Persistent drainage suggests communication with the pancreatic duct, which can be confirmed by a contrast medium catheter study or by an amylase estimation of the drainage fluid (amylase concentrations >100000 units/litre strongly suggest communication). In this case, the catheter should be left until an established fistulous track has formed. If a pseudocystogastrostomy has been performed, a double pigtail stent may be inserted across the pseudocyst and stomach lumen to establish internal drainage. The stent may be removed endoscopically later.

The reported success rate of percutaneous pseudocyst drainage ranges between 60 and 90%.

PANCREATIC PHLEGMON, ABSCESS AND NECROSIS

Intervention, when required, is primarily surgical. Percutaneous drainage does not usually succeed because of the extensive debris and necrotic tissue present. However, radiological intervention may be useful in two situations. Radiologically guided

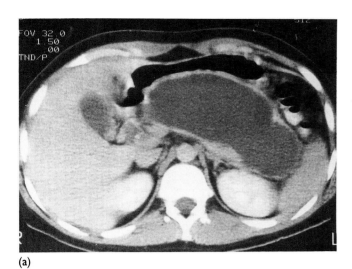

(a)

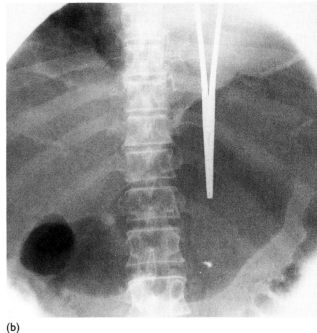

(b)

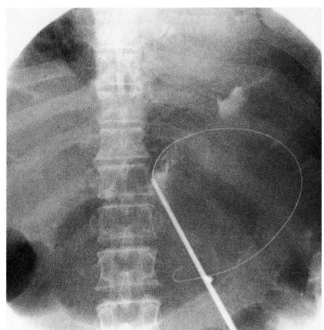

(c)

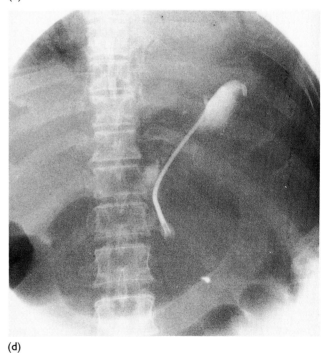

(d)

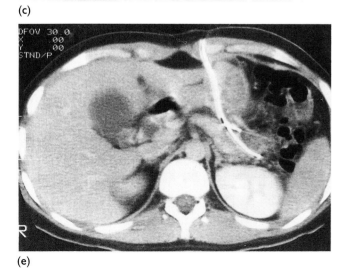

(e)

Fig. 7.18 *Transgastric pseudocyst drainage*
(a) Computed tomographic (CT) scan shows a large pseudocyst displacing and compressing the stomach anteriorly. (b) The forceps tip marks the proposed skin entry point. (c) The pseudocyst has been punctured, a wire introduced and a drainage catheter advanced after appropriate track dilation. (d) The final position of the catheter. (e) A subsequent CT scan (not shown) showed a residual cavity. The drainage catheter was exchanged for a larger one and repositioned. This CT scan shows resolution of the pseudocyst. Note that the drainage catheter traverses the length of the cavity, which has allowed complete drainage.

diagnostic needle aspiration can determine whether a pancreatic phlegmon is infected, which necessitates more aggressive therapy. Radiological drainage may also be used as an interim measure to improve the condition of a critically ill patient before surgical debridement and drainage. Some radiologists believe that percutaneous catheter drainage should be tried in all cases even though surgery will eventually be needed.

Pelvic collections

Deep pelvic collections are generally regarded as inaccessible to radiological drainage because they are surrounded by bowel, pelvic organs and bone. However, several access routes are available for radiological drainage depending on the location of the collection (Fig. 7.19). A CT scan is usually required for precise localization before drainage.

PRECOCCYGEAL/RETRO-ANAL APPROACH

Low presacral collections are readily drained by this approach using fluoroscopic guidance and placing the patient in the left lateral position.

TRANSRECTAL AND TRANSVAGINAL APPROACHES

Pelvic collections close to the anterior wall of the rectum are suitable for transrectal or transvaginal drainage (Mauro *et al.*, 1985; Nosher *et al.*, 1987).

Transrectal drainage is performed using fluoroscopic guidance with the patient in the left lateral position. No colonic preparation is required. A digital rectal examination is performed before the procedure to locate the bulging abscess. At our institution, a plastic barium enema tube preloaded with an 18 G Teflon-sheathed needle is inserted into the rectum and manoeuvred into the appropriate position (Fig. 7.20). The bulging abscess is usually well visualized fluoroscopically as it indents the anterior rectal wall. If required, air or water-soluble contrast medium may be injected through the side arm of the enema tube to outline the rectum. The collection is punctured and a guidewire inserted. The track is dilated serially and a 10–14 Fr drainage catheter inserted (Fig. 7.21). The procedure is surprisingly pain free.

Transvaginal drainage is suitable in multiparous women (in nulliparous women this approach can be quite painful) and is performed using a combination of ultrasound and fluoroscopic guidance. The ultrasound transducer is placed on the patient's lower abdomen and a longitudinal scan outlining the abscess and vaginal vault is obtained (Fig. 7.22). The patient is placed in a lithotomy position and a self-retaining vaginal speculum is inserted. The vagina is thoroughly cleansed with Betadine (povidone iodine; Purdue Frederick, Norwalk, CT). Drainage is performed either in a single step with a trocar catheter or using the Seldinger technique. Correct positioning of the needle/drainage catheter for entry into the collection is confirmed sonographically by

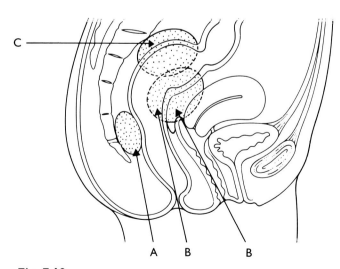

Fig. 7.19
Access routes to pelvic collections. Precoccygeal collections are amenable to a retro-anal approach (A). Collections in the pouch of Douglas or prerectal region (B) are amenable to transrectal or transvaginal approaches. Higher pelvic collections not amenable to these two approaches may be suitable for transgluteal drainage (C).

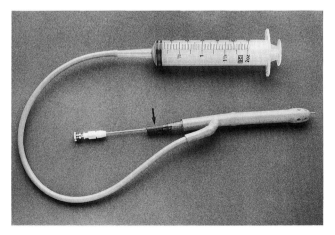

Fig. 7.20
A barium enema tube preloaded with an 18 G Teflon-sheathed needle. A tight seal between the needle and the tube end is made by plugging the tip of a 24 Fr red rubber catheter into the tube end (arrow). Air or contrast medium may be injected through the tube side arm.

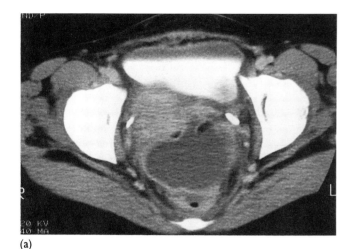

(a)

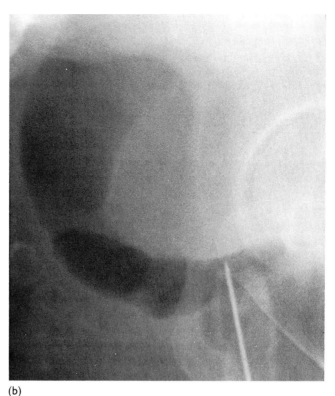

(b)

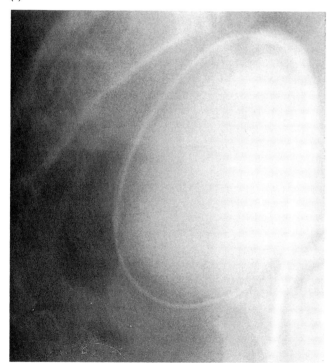

(c)

Fig. 7.21 *Transrectal approach*
(a) Computed tomographic (CT) scan shows a large collection anterior to the rectum. (b) Lateral radiograph shows the collection indenting the air-filled rectum. The barium enema tube tip with needle has been positioned for puncture. (c) The collection has been entered, contrast medium injected, a wire inserted and the drainage catheter placed.

distortion of the posterior vaginal vault as the needle is advanced. Like transrectal drainage, transvaginal drainage is relatively pain free.

Imaging for transvaginal drainage may also be accomplished via an endovaginal probe (Van-Derkolk, 1991; van Sonnenberg *et al.*, 1991). Computed tomographic-guided transrectal drainage has also been described (Gazelle *et al.*, 1991).

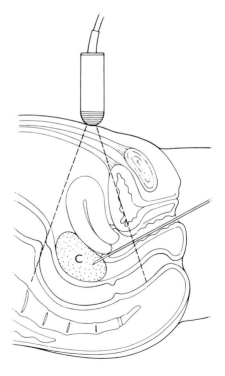

Fig. 7.22
Schematic diagram of transvaginal drainage. A longitudinal scan of the pelvis is obtained by placing the ultrasound transducer in the suprapubic region. The collection (C) is entered through a posterior forniceal puncture.

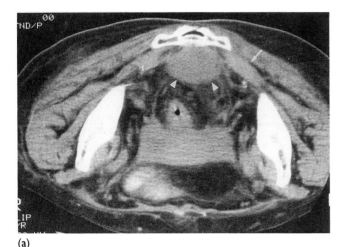

(a)

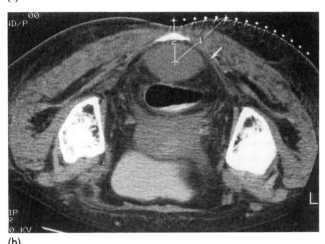

(b)

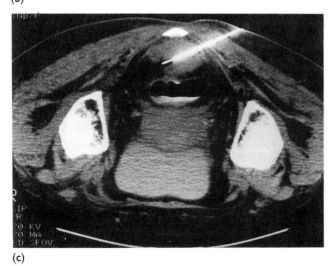

(c)

Fig. 7.23 *Transgluteal drainage*
(a) A computed tomographic (CT) scan of the pelvis with the patient prone shows a presacral collection (arrowheads). Note the piriformis muscle (arrow). (b) The skin entry point and needle trajectory are chosen using the skin marker grid. Note that the level chosen for drainage is below the piriformis muscle through the sacrospinous ligament (arrow) and medial (i.e. adjacent to the sacrum). (c) A CT scan of the needle in the collection, confirming accurate positioning.

TRANSGLUTEAL APPROACH

The transgluteal approach (Butch *et al*, 1986) is through the greater sciatic foramen and requires CT guidance because ultrasound imaging is poor in this region. A good understanding of the anatomic structures in the area is mandatory for safe drainage. The piriformis muscle passes through the centre of the foramen, while the sacral plexus, superior and inferior gluteal vessels and superior gluteal nerve cross the foramen in its superior aspect. Inferiorly, the sacral plexus continues as the sciatic nerve, and crosses the anterior third of the foramen. Catheter drainage should be performed through the inferomedial portion of the greater sciatic foramen just lateral to the sacral bone at the level of the sacrospinous ligament (Fig. 7.23).

REFERENCES

ALLISON DJ, ADAM A (1988). Percutaneous liver biopsy and track embolization with steel coils. *Radiology* **169**: 261–3.

ANDRIOLE JG, HAAGA JR, ADAMS RB, NUNEZ C (1983). Biopsy needle characteristics assessed in the laboratory. *Radiology* **148**: 659–62.

AXEL L (1984). Simple method for performing oblique CT-guided needle biopsies. *American Journal of Roentgenology* **143**: 341–2.

BISSONNETTE RT, GIBNEY RJ, BERRY BR, BUCKLEY AR (1990). Fatal carcinoid crisis after percutaneous fine-needle biopsy of hepatic metastasis: case report and literature review. *Radiology* **174**: 751–2.

BUTCH RJ, MUELLER PR, FERRUCCI JT JR, WITTENBERG J, SIMEONE JF, WHITE EM, BROWN AS (1986). Drainage of pelvic abscesses through the greater sciatic foramen. *Radiology* **158**: 487–91.

FAGELMAN D, CHESS Q (1990). Nonaspiration fine-needle cytology of the liver: A new technique for obtaining diagnostic samples. *American Journal of Roentgenology* **155**: 1217–19.

FERRUCCI JT, WITTENBERG J, MUELLER PR, SIMEONE JF, HARBIN WP, KIRKPATRICK RH, TAFT PD (1980). Diagnosis of abdominal malignancy by radiologic fine-needle aspiration biopsy. *American Journal of Roentgenology* **134**: 323–30.

GAMBLE P, COLAPINTO RF, STRONELL RD, COLMAN JC, BLENDIS L (1985). Transjugular liver biopsy: A review of 461 biopsies. *Radiology* **157**: 589–93.

GAZELLE GS, HAAGA JR, STELLATO TA, GAURDERER MWL, PLECHA DT (1991). Pelvis abscesses: CT guided transrectal drainage. *Radiology* **181**: 49–51.

GERZOF SG, ROBBINS AH, JOHNSON WC, BIRKETT D, NABSETH D (1981). Percutaneous catheter drainage of abdominal abscesses: A five-year experience. *New England Journal of Medicine* **305**: 653–7.

GOBIEN RP, STANLEY JH, SCHABEL SI, CURRY NS, GOBIEN BS,

VUJIC I, REINES HD (1985). The effect of drainage tube size on adequacy of percutaneous abscess drainage. *Cardiovascular and Interventional and Radiology* **8**: 100–2.

HAAGA JR, LIPUMA JP, BRYAN PJ, BALSARA BJ, COHEN AM (1983). Clinical comparison of small- and large-caliber cutting needles for biopsy. *Radiology* **146**: 665–7.

HAWKINS IF, AKINS EW, MLADINICH C, TUPLER R, SIRAGUSA RJ (1988). Transvisceral access using a blunt needle: Technical note. *Seminars in Interventional Radiology* **5**(2): 149–51.

HEUFTLE MG, HAAGA JR (1986). Effect of suction on biopsy sample size. *American Journal of Roentgenology* **147**: 1014–16.

HORTON JA, BANK WO, KERBER CW (1980). Guiding the thin spinal needle. *American Journal of Roentgenology* **134**: 845–6.

LANG EK, SPRINGER RM, GLORIOSA LW III, CAMMARATA CA (1986). Abdominal abscess drainage under radiologic guidance: Causes of failure. *Radiology* **159**: 329–36.

LEE TG, KNOCHEL JQ (1982). Air as an ultrasound contrast marker for accurate determination of needle placement: Tumour biopsy localization and other applications. *Radiology* **143**: 787–8.

LEES WR, HALL-CRAGGS MA, MANHIRE AR (1985). Five years' experience of fine-needle aspiration biopsy: 454 consecutive cases. *Clinical Radiology* **36**: 517–20.

LINDGREN PG (1982). Percutaneous needle biopsy: A new technique. *Acta Scandinavica Radiologica (Diagnostica)* **23**: 653–6.

MATALON TAS, SILVER B (1990). US guidance of interventional procedures. *Radiology* **174**: 43–7.

MATZINGER FRK, HO C-S, YEE AC, GRAY RR (1988). Pancreatic pseudocysts drained through a percutaneous transgastric approach: Further experience. *Radiology* **167**: 431–4.

MAURO MA, JAQUES PF, MANDELL VS, MANDEL SR (1985). Pelvic abscess drainage by the transrectal catheter approach in men. *American Journal of Roentgenology* **144**: 477–9.

McGAHAN JP (1984). Percutaneous biopsy and drainage procedures in the abdomen using a modified coaxial technique. *Radiology* **153**: 257–8.

MUELLER PR, VAN SONNENBERG E, FERRUCCI JT JR (1984). Percutaneous drainage of 250 abdominal abscesses and fluid collections. Part II: Current procedural concepts. *Radiology* **151**: 343–7.

MUELLER PR, FERRUCCI JT, BUTCH RJ, SIMEONE JF, WITTENBERG J (1985). Inadvertent percutaneous catheter gastroenterostomy during abscess drainage: Significance and management. *American Journal of Roentgenology* **145**: 387–91.

NOSHER JL, WINCHMAN HK, NEEDELL GS (1987). Transvaginal pelvic abscess drainage with US guidance. *Radiology* **165**: 872–3.

PARKER SH, HOPPER KD, YAKES WF, GIBSON MD, OWNBEY JL, CARTER TE (1989). Image-directed percutaneous biopsies with a biopsy gun. *Radiology* **171**: 663–9.

RILEY SA, ELLIS WR, IRVING HC, LINTOTT DJ, AXON ATR,

LOSAUSKY MS (1984). Percutaneous liver biopsy with plugging of needle track: A safe method for use in patients with impaired coagulation. *Lancet* **ii**: 436.

SOLBIATI L, LIVRAGHI T, DE PRA L, TERRACE T, MASCIADRI N, RAVETTO C (1985). Fine needle biopsy of hepatic hemangioma with sonographic guidance. *American Journal of Roentgenology* **144**: 471–4.

VAN DERKOLK HL (1991). Small, deep pelvic abscesses: Definition and drainage guided with an endovaginal probe. *Radiology* **181**: 283–4.

VAN SONNENBERG E, CASOLA G, VARNEY RR, WITTICH GR (1989). Imaging and interventional radiology for pancreatitis and its complications. *Radiologic Clinics of North America* **27**: 65–72.

VAN SONNENBERG E, D'AGOSTINO HB, CASOLA G, GOODACRE BW, SANCHEZ RB, TAYLOR B (1991). US-guided transvaginal drainage of pelvic abscesses and fluid collections *Radiology* **181**: 53–6.

VAN SONNENBERG E, MUELLER PR, FERRUCCI JT JR (1984). Percutaneous drainage of 250 abdominal abscesses and fluid collections. Part I: Results, failures, and complications. *Radiology* **151**: 337–41.

VAN SONNENBERG E, MUELLER PR, FERRUCCI JT JR, NEFF CC, SIMEONE JF, WITTENBERG J (1982). Sump catheter for percutaneous abscess and fluid drainage by trocar or Seldinger technique. *American Journal of Roentgenology* **139**: 613–14.

VAN SONNENBERG E, WITTENBERG J, FERRUCCI JT, MUELLER PR, SIMEONE JF (1981). Triangulation method for percutaneous needle guidance: The angled approach to upper abdominal masses. *American Journal of Roentgenology* **137**: 757–61.

VOGELZANG RL, TOBIN RS, BURSTEIN S, ANSCHUETZ SL, MARZANO M, KOZLOWSKI JM (1987). Transcatheter intracavitary fibrinolysis of infected extravascular hematomas. *American Journal of Roentgenology* **148**: 378–80.

WELCH TJ, SHEEDY PF, JOHNSON CD, JOHNSON CM, STEPHENS DH (1989). CT-guided biopsy: Prospective analysis of 1000 procedures. *Radiology* **171**: 493–6.

WITTENBERG J, MUELLER PR, FERRUCCI JT, SIMEONE JF, VAN SONNENBERG E, NEFF CC, PALERMO RA, ISLER RJ (1982). Percutaneous core biopsy of abdominal tumors using 22 gauge needles: Further observations. *American Journal of Roentgenology* **139**: 75–80.

YEUNG EY, THURSTON W, QUIGLEY MJ, HO C-S (1990). US guidance of interventional procedures. *Radiology* **176**: 289–90 (letter).

YUEH N, HALVORSEN RA, LETOURNEAU JG, CRESS JR (1989). Gantry tilt technique for CT-guided biopsy and drainage. *Journal of Computer Assisted Tomography* **13**: 182–4.

I would like to thank Dr Siok Min Tan for her invaluable comments and Ms Edvige Coretti for her expert secretarial assistance in the preparation of the manuscript.

CHAPTER 8

Liver biopsy in patients with abnormal coagulation: alternatives to the transjugular approach

Andy Adam

Transjugular liver biopsy   119

Plugged liver biopsy   120

Percutaneous transcaval tumour biopsy   124

References   126

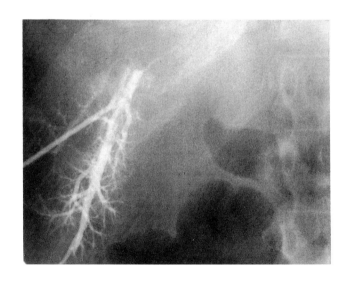

Percutaneous liver biopsy is an established procedure with a very low morbidity and mortality in patients with normal coagulation. However, many patients with diffuse liver diseases such as cirrhosis or widespread malignant infiltration have abnormalities of blood coagulation which increase significantly the risk of bleeding after a biopsy. An imaging-targeted approach is usually unnecessary in these patients as most of the liver is affected by the disease process and a 'blind' biopsy is likely to produce useful histological information. The main problem to be overcome is the risk of bleeding and the traditional way of performing a biopsy in this situation is via the transjugular approach. This method is described only in outline below as its popularity is declining following the development of less cumbersome methods in recent years.

TRANSJUGULAR LIVER BIOPSY

Technique

The biopsy is performed with a needle reaching the liver parenchyma from inside the vascular tree thus avoiding external bleeding. Either internal jugular vein may be used. The patient is placed in the supine position with the head turned away from the side to be punctured and the catheter is introduced under local anaesthesia using the Seldinger technique. The internal jugular vein lies posterior and lateral to the common carotid artery within the carotid sheath. The puncture is made approximately halfway between the angle of the jaw and the clavicle, just anterior to the sternocleidomastoid muscle. Any needle that will accept a 0.038 inch guidewire may be used for the venous puncture. Following insertion of the guidewire a 9 Fr dilator is inserted into the vein followed by the 9 Fr biopsy catheter which is advanced, under fluoroscopic control, via the inferior vena cava to the right hepatic vein (Fig. 8.1). The biopsy needle is then advanced until it lies at the catheter tip. A 10 ml syringe containing saline is attached to the hub of the needle and the catheter is wedged into the hepatic vein. The patient is asked to hold his breath in maximum inspiration, following which the needle is inserted into the liver

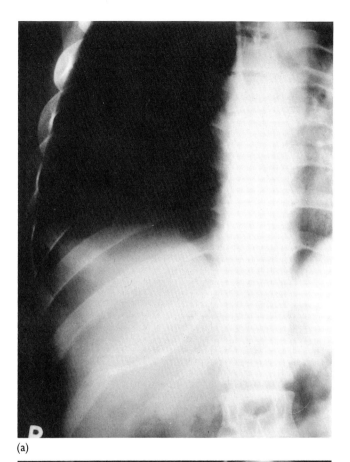

(a)

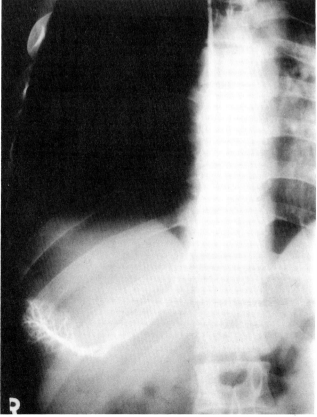

(b)

Fig. 8.1
(a) Transjugular catheter positioned in the right hepatic vein.
(b) The catheter is shown in a wedged position.

parenchyma. The biopsy sample is obtained using the Menghini '1 second' technique. The needle is plunged into the liver substance and removed in one continuous motion while suction is maintained with the 10 ml syringe. The inner stylet of the biopsy needle prevents the aspiration of the specimen into the syringe.

A variation of the above technique involves obtaining the biopsy with the catheter in a free, rather than a wedged, position within the hepatic vein. This is said to minimize the risk of perforation of the liver capsule (Gamble *et al*, 1985).

Complications

The main complication is perforation of the liver capsule leading to intraperitoneal bleeding. Capsular perforation is seen in approximately 3.9% of patients (Gamble *et al*., 1985), but usually there are no clinical sequelae. Clinically significant bleeding has been reported in only eight (0.35%) of 2271 transjugular biopsies reported to date (Colapinto, 1985). The mortality rate is approximately 0.13% (Colapinto, 1985).

Haematomas at the puncture site are common but usually insignificant. Transient hoarseness or Horner's syndrome may occur and are probably related to the local anaesthetic.

Fever and rigors sometimes occur in the 24 hours following the procedure. The incidence may be decreased by ultrasonic cleaning of the biopsy needle. The presence of the catheter in the right atrium sometimes results in transient arrhythmias and electrocardiographic (ECG) monitoring is usually employed during the procedure.

Results

It is possible to pass the needle into the liver in over 90% of cases. Difficulty may be encountered if the liver is small and the diaphragm elevated by ascites, in which case the right hepatic vein runs more horizontally, forming an angle with the inferior vena cava that is difficult or impossible to negotiate with a stiff needle. More commonly, failure occurs because the specimen is too small or too fragmented to allow accurate pathological diagnosis. This occurs in 15–20% of cases. Probably the most important factor in the success of transjugular liver biopsy is the presence of a highly trained pathologist who is able to make an accurate diagnosis from a relatively minute particle of tissue.

Notwithstanding the problems associated with transjugular biopsy there is no doubt that it is a very safe technique in experienced hands and when it is employed in large hospitals where the necessary pathological expertise exists the diagnostic yield is very high. However, few hospitals combine the radiological and histological skills necessary for a high success rate. In addition the procedure is cumbersome and time consuming and is often frightening for the patient. It is these considerations that have lead to the development of the alternative methods described below.

PLUGGED LIVER BIOPSY

There are various versions of this technique but the basic principle underlying it is the performance of the biopsy through a sheath followed by embolization of the biopsy track. The needle used is usually of the 'Trucut' type which has been employed for many years for the performance of liver biopsy and has been shown to produce excellent histological specimens. This type of needle compares very favourably with the instrument employed for transjugular biopsy in the size and quality of the tissue particles obtained. The first report of plugged liver biopsy was by Riley *et al*. (1984). However, the authors of that paper generously acknowledged that 'the basic principle of performing a conventional liver biopsy with plugging of the needle track was the idea of Dr D.J. Irving of Lewisham Hospital'.

Technique

The method described here is that employed in our unit. Variations are mentioned briefly below.

An intravenous line is introduced and the patient is given platelets and/or coagulation factors as necessary. Mild sedation is occasionally employed although high doses of narcotics are potentially dangerous in patients with abnormal liver function and, in most cases, are unnecessary.

The biopsy is performed in the angiography unit using a specially designed biopsy and embolization kit (Fig. 8.2) (William Cook Europe, Bjaeverskov, Denmark). Preliminary fluoroscopic screening helps to identify the position of the liver and any gas in adjacent bowel. The needle used is 14 Fr in calibre, 20 cm long and has a 'Trucut' action. The 8 Fr sheath is placed over the closed needle which is then advanced 4–5 cm into the liver carrying the sheath

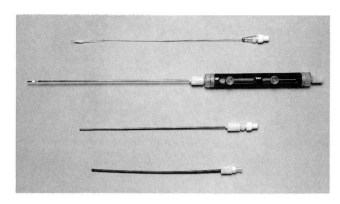

Fig. 8.2
Biopsy kit comprises the special purpose needle, the sheath, a preloaded coil delivery needle and a plastic obturator.

with it. The needle is introduced parallel to the table top at an angle of approximately 30° cephalad in an attempt to avoid the gall bladder and major vascular structures. Once inside the liver parenchyma the needle is opened, advanced beyond the sheath tip another 2–3 cm into the liver and closed again. After the needle with the tissue fragment is withdrawn, there is often profuse bleeding from the sheath. Contrast medium injections usually show that this is because the biopsy track communicates with a portal or hepatic vein radicle. After accurate demonstration of such a communication, a specially designed needle (20 gauge, 28.6 cm cannula with a 0.24 mm inner core) carrying a preloaded embolization coil (Fig. 8.3) is inserted into the sheath, and the coil is

deposited in the biopsy track close to the point of communication between the track and the damaged vein.

The above method is outlined in Fig. 8.4 and an example is shown in Fig. 8.5. Following deposition of the coil haemostasis usually results within 1–2 min and on gentle injection of contrast medium through the sheath the vessel responsible for the bleeding is no longer visible. If bleeding continues more coils can be deposited. Occasionally it may be advantageous to combine coil embolization with the injection of particulate embolic materials such as Ivalon.

Following embolization of the track at the point of communication with the vessel responsible for the bleeding the sheath is gradually withdrawn and more contrast medium is injected to demonstrate any other communications between the track and hepatic blood vessels. These communications can then be embolized in the same way (Fig. 8.6). If a coil is deposited very close to the liver capsule and its position needs to be adjusted a 6 Fr plastic obturator, which is contained in the kit, is used to position the coil optimally within the track.

The technique described above not only ensures the immediate recognition of any bleeding but, by demonstrating the exact location of any vascular communication with the track allows the precise positioning of embolization coils at the most effective site. It is possible to introduce embolization coils without the use of the preloaded needle described either by using a conventional distal delivery guidewire or by simply pushing the coils through

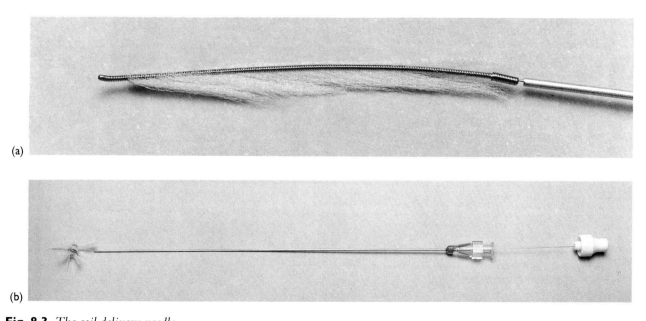

(a)

(b)

Fig. 8.3 *The coil delivery needle*
(a) Detail of the needle tip showing the coil straightened by the inner core of the needle. (b) The inner core has been withdrawn allowing the coil to be delivered.

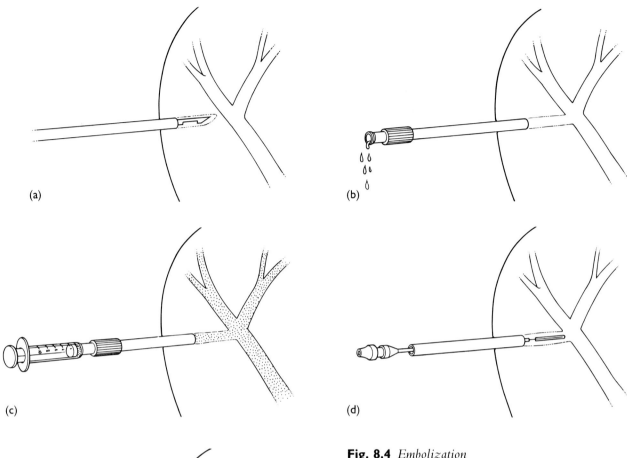

(a)

(b)

(c)

(d)

(e)

Fig. 8.4 *Embolization method*
(a) The biopsy needle is inserted through the sheath. (b) After withdrawal of the needle, bleeding usually occurs. This is typically due to communication of the track with a portal venous radicle. (c) Injection of contrast medium usually demonstrates communication of the biopsy track with the portal venous system. (d) The needle carrying a preloaded coil is inserted into the track. (e) The track is embolized with steel coils. Reproduced from Allison and Adam (1988) with permission from *Radiology*.

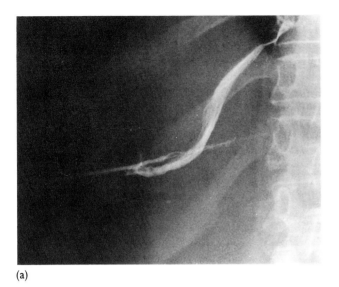

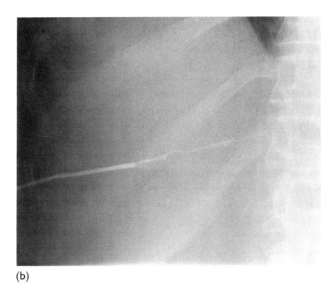

(a)   (b)

Fig. 8.5
(a) A postbiopsy contrast material-enhanced study shows communication of the biopsy track with a hepatic vein. (b) The abnormal communication has been successfully occluded with the percutaneous coil delivery system. The coil can be seen in the biopsy track, and the injected contrast medium is now confined to the sheath. Reproduced from Allison and Adam (1988) with permission from *Radiology*.

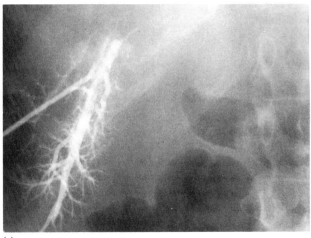

(a)

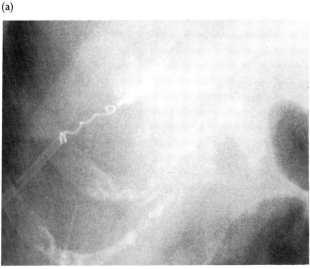

(c)

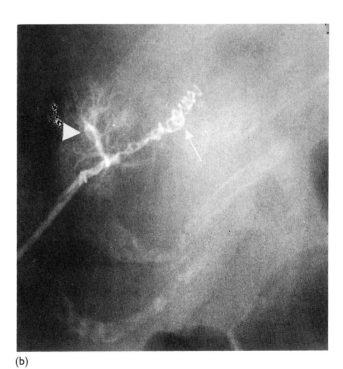

(b)

Fig. 8.6
(a) The biopsy has produced a communication with a large portal radicle, as demonstrated by this contrast-enhanced image. (b) After successful occlusion of the communication with percutaneous coils (white arrow), a second vascular communication (arrowhead) is seen on partial withdrawal of the sheath and a further injection of contrast medium. This communication has been created by the passage of the needle and sheath and is unrelated to the site of actual biopsy. The coil has been inserted across the origin of this communication. (c) A few minutes later thrombosis of the track is complete, and the sheath may be safely withdrawn. Reproduced from Allison and Adam (1988) with permission from *Radiology*.

a catheter introduced into the biopsy sheath. However, both of these methods are more time consuming and cumbersome than the present technique, and both are much more difficult for one person alone to perform. The technique described allows the accurate and permanent placement of a coil at any desired location in the track without the risk of either outward displacement, causing bleeding into the peritoneum, or inward displacement, causing pulmonary embolization.

Variations of the plugged biopsy technique

The concept of embolization of a biopsy track through a sheath is not new; interventional radiologists have used the technique for many years with varying degrees of success (Riley *et al.*, 1984; Allison, 1986; Chuang and Alspaugh, 1988). Most methods have employed particulate materials such as Gelfoam (gelatine) fragments as the embolic agent. Although in many cases this practice is successful, it is not a suitable method for stopping massive bleeding, such as may occur from a transected artery or large portal radicle, because the embolic fragments may be swept out of the track by the force of the blood flow. The technique also suffers from the disadvantage that embolic materials may be injected along the biopsy track into the portal system, an event that is clearly undesirable in a patient whose liver function may already be seriously compromised.

Crummy *et al.*, (1989) have suggested that embolization of biopsy tracks can be performed without the use of an outer sheath, simply by releasing the inner stylet of a 'Trucut' needle from the cannula handle and removing it. The outer (cannula) portion can then be used to inject contrast media, steel coils or gelatine sponge plugs to obtain haemostasis.

Gazelle *et al.* (1990) have described a haemostatic protein/polymer sheath (PPS) for plugged biopsy. In its dry state the PPS is extremely rigid yet rapidly softens and expands when it comes into contact with liquid at body temperature. This sheath can be inserted through the cannula of a 'Trucut' needle to achieve haemostasis after a biopsy. This is a promising method which proved rapid and effective experimentally in pigs, but the sheath is not yet commercially available.

The technique described above which employs steel coils has been modified recently for use in conjunction with the 'Biopty' gun (Dawson *et al.*, 1992). The latter has proved extremely effective at producing excellent histological specimens using a relatively small calibre needle by its rapid, controlled spring-powered delivery mechanism. Eighteen gauge 'Trucut' style needles can be used with a Biopty gun and a purpose-built 18 gauge sheath is available for the plugged biopsy procedure (William Cook Europe, Bjaeverskor, Denmark). The smaller needles are intrinsically safer than the standard 14 gauge needle and are, importantly, quite flexible. A modified version of the 20 gauge embolization needle described above is used to deposit a coil in the track immediately adjacent to the bleeding vessel. The embolization needle assembly is 1 cm longer than the previously commercially produced device because the needle employed with the Biopty gun is longer than the one included in the biopsy and embolization set described previously.

PERCUTANEOUS TRANSCAVAL TUMOUR BIOPSY

Occasionally primary or metastatic malignant liver tumours invade the intrahepatic portion of the inferior vena cava. The diagnosis is often made by ultrasound scanning preferably employing a Doppler technique. This situation offers the radiologist the opportunity to obtain a biopsy specimen through the inferior vena cava using an approach via one of the femoral veins (O'Donnell and Adam, 1988). An approach from above via internal jugular vein and the superior vena cava is best avoided because of the risk of tumour embolization of the pulmonary arteries. There are various versions of this technique but our preference is for one which employs a modified cardiac bioptome inserted through a long 8 Fr sheath (Fig. 8.7). The bioptome has stainless steel jaws controlled by an inner drive wire attached to a proximal control handle. The forceps produce consistently good, histologically useful pieces of tissue.

Following puncture of a femoral vein and the performance of intravenous cavography using standard methods (Fig. 8.8a), the sheath is advanced into the inferior vena cava and its tip is positioned immediately below the level of obstruction or stenosis by tumour. The bioptome is introduced into the sheath and advanced with its jaws closed until it is well wedged into the tumour (Fig. 8.8b). The jaws are then opened (Fig. 8.8c). The instrument is pushed into the tumour a little further, the jaws are closed and the bioptome is withdrawn.

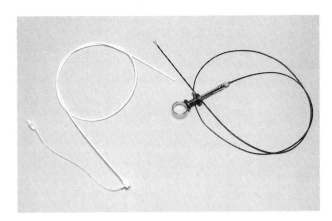

Fig. 8.7 *The bioptome instrument and its long 8 Fr sheath*

A modification of this method which is very useful if the vena cava is not completely obstructed involves the use of the 'roadmap' facility available on some digital subtraction angiographic equipment. This allows continuous visualization of the site of the tumour thrombus during repeated biopsies (Jackson and Adam, 1992).

Transcaval biopsy consistently produces excellent histological specimens.

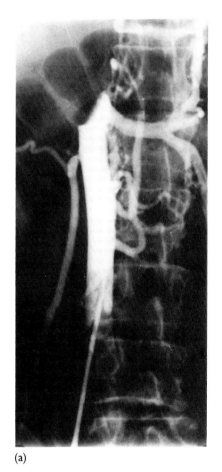

(a)

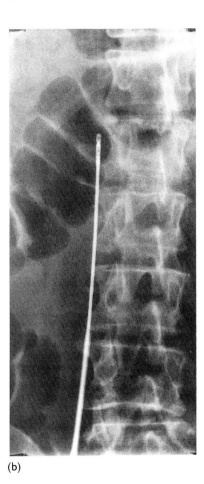

(b)

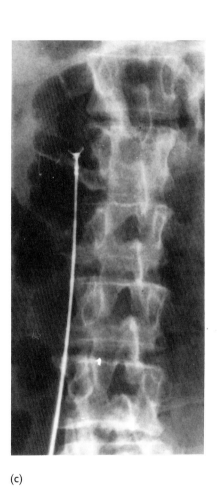

(c)

Fig. 8.8
(a) Vena cavography via the right femoral vein demonstrates occlusion of the inferior vena cava. Multiple collateral veins are also seen.
(b) The bioptome is shown with its jaws closed, wedged into the tumour. (c) The jaws of the instrument are open immediately prior to taking the biopsy. Reproduced from O'Donnell and Adam (1988) with permission from the *Journal of Interventional Radiology*.

REFERENCES

ALLISON DJ (1986). Interventional radiology. In: Diagnostic Radiology: An Anglo-American Textbook of Imaging, pp. 2121–65. Grainger RG, Allison DJ eds. Churchill Livingstone, Edinburgh.

ALLISON DJ, ADAM A (1988). Percutaneous liver biopsy and track embolization with steel coils. *Radiology* **169**: 261–3.

CHUANG VP, ALSPAUGH JP (1988). Sheath needle for liver biopsy in high-risk patients. *Radiology* **166**: 261–2.

COLAPINTO RF (1985). Transjugular biopsy of the liver. *Clinics in Gastroenterology* **14**: 451–67.

CRUMMY AB, McDERMOTT JC, WOJTOWYCZ m (1989). A technique for embolization of biopsy tracts. *American Journal of Radiology* **153**: 67–8.

DAWSON P, ADAM A, EDWARDS R (1992). Technique for steel coil embolization of liver biopsy track for use with the 'Biopty' needle. *British Journal of Radiology* **65**: 538–540.

O'DONNELL C, ADAM A (1988). Percutaneous transcaval tumour biopsy using bioptome forceps. *Journal of Interventional Radiology* **3**: 19–21.

GAMBLE P, COLAPINTO RF, STRONELL RD, COLMAN JC, BLENDIS L (1985). Transjugular liver biopsy: a review of 461 biopsies. *Radiology* **157**: 589–93.

GAZELLE GS, HAAGA JR, NEUHANSER D (1990). Hemostatic protein–polymer sheath: new method to enhance hemostasis at percutaneous biopsy. *Radiology* **175**: 671–4.

JACKSON JE, ADAM A (1991). Percutaneous transcaval tumour biopsy using a 'road map' technique. *Clinical Radiology* **44**: 195–196.

RILEY SA, IRVING HC, AXON ATR, ELLIS WR, LINTOTT DJ, LOSOWSKY MS (1984). Percutaneous needle with plugging of needle track: a safe method for use in patients with impaired coagulation. *Lancet* **ii**: 436.

3

OESOPHAGEAL AND GASTROINTESTINAL INTERVENTION

Oesophageal strictures

Steven G. Meranze and Gordon K. McLean

Overview and indications for radiological methods   129

Contraindications   135

Instrumentation   135

Technique   136

Common mistakes and how to avoid them   137

Complications   138

Postprocedural care   138

Results   138

Conclusion   139

References   139

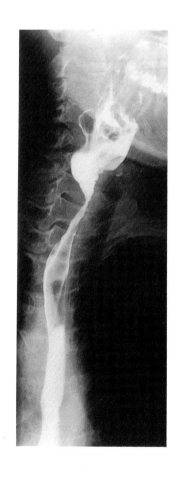

Over the past decade, the application of radiologically guided balloon dilation techniques has allowed the interventional radiologist to become an active participant in the treatment of enteric strictures, often in a safer and more controlled fashion than would be possible with alternative techniques such as surgery or bougienage (London *et al.*, 1981). The following discussion reviews the fundamentals of interventional radiologic techniques in the clinical management of oesophageal strictures.

OVERVIEW AND INDICATIONS FOR RADIOLOGICAL METHODS

Aetiology

The majority of oesophageal strictures treated in most practices are associated with reflux oesophagitis (Figs 9.1 and 9.2). These are often fairly short in length and are most frequently located in the distal portion of the oesophagus, unless they are associated with Barrett's oesophagus, in which case they may occur more proximally. Reflux strictures are commonly treated with progressive bougienage by either the patient or by an endoscopist. Reported success rates exceed 80% in some series (Wesdorp *et al.*, 1982). Some lesions, because of either their severity or location, cannot be approached as safely or effectively with bougienage. These are frequently the type of lesions referred for fluoroscopically guided procedures.

Ingestion of a caustic substance (Fig. 9.3) often results in longer, more irregular, lesions which may be distributed throughout the oesophagus. These lesions occur most frequently in areas of pooling or stasis, such as the oesophagogastric junction. They may be more susceptible to injury during dilation due to the friability of their mucosa, and treatment should be postponed until scarring has developed. Additional benign causes of strictures which have been treated with dilation techniques with varying

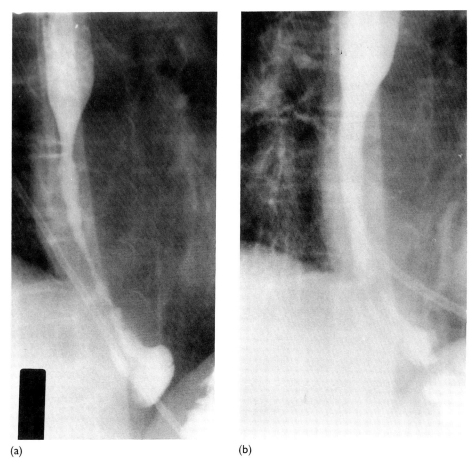

(a)   (b)

Fig. 9.1
A patient with a typical reflux stricture, status postNissen fundoplication. Attempts to treat this lesion endoscopically had been unsuccessful. (a) A barium swallow demonstrates a very high grade stricture in the distal oesophagus. (b) Following successful dilation to 9 mm there is a marked improvement in lumen diameter.

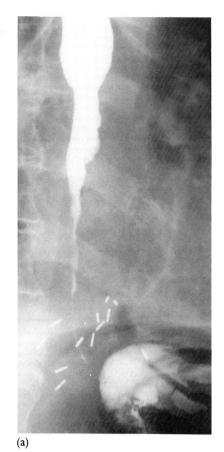

(a)

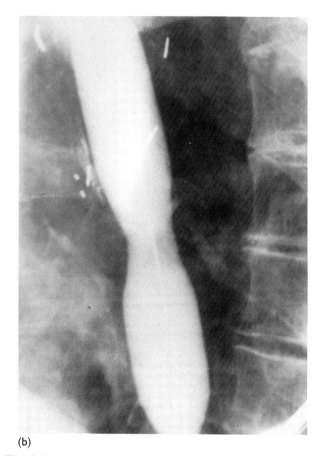

(b)

Fig. 9.2
Sequence illustrating successful dilation of a severe reflux stricture in a patient status post antireflux surgery. (a) A barium swallow demonstrating an almost complete obstruction in the distal oesophagus. (b) Partially inflated balloon shows minimal 'waist'. (c) Postprocedure film. Note the mucosal irregularity. The patient experienced symptomatic relief and later films demonstrated a smooth lumen.

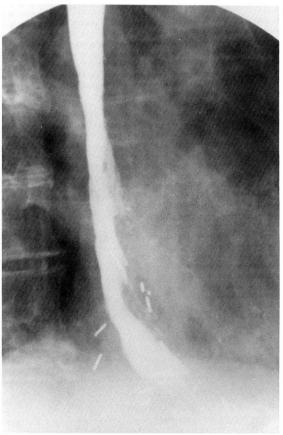

(c)

Fig. 9.3
Sequence demonstrating the treatment of a stricture following ingestion of a caustic substance. (a) Baseline barium swallow demonstrating proximal stricture. (b–e) Sequence demonstrating progressive dilation to 15 mm. (f) Postprocedure film demonstrating marked improvement.

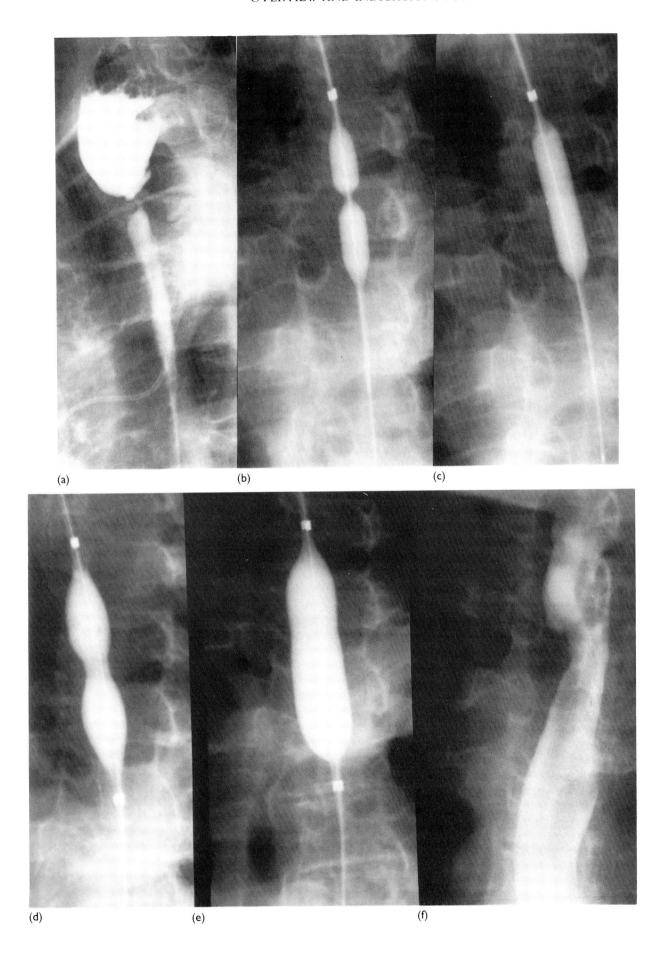

(a)

(b)

(c)

(d)

(e)

(f)

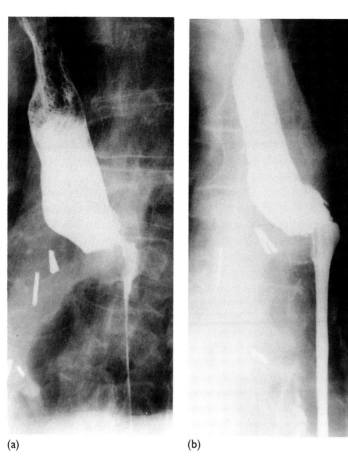

(a)   (b)

Fig. 9.4
Successful dilation of an anastomotic stricture which developed following oesophagogastrectomy for malignancy. (a) Baseline film at the time the patient presented with severe dysphagia.
(b) Following successful dilation to 12 mm.

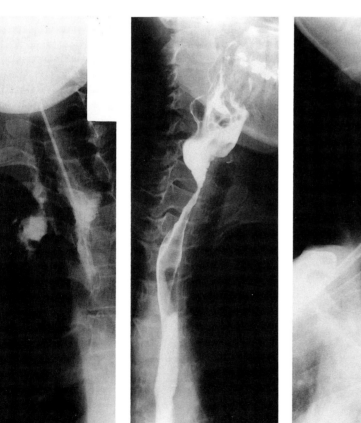
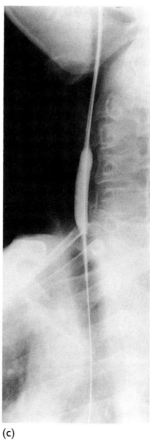

(a)   (b)   (c)

Fig. 9.5
This patient developed an oesophageal perforation at the time of an attempted endoscopic dilation of a midoesophageal stricture.
(a) Baseline film demonstrating site of perforation which was initially managed with a drainage procedure. (b) A later barium swallow demonstrates that a stricture developed at the site of the perforation.
(c) The stricture was easily traversed and successfully dilated.

degrees of success include achalasia as well as oesophageal webs (Webb *et al.*, 1984). Treatment of achalasia necessitates disruption of the muscle fibres of the lower oesophageal sphincter, and large bougies or multiple balloons may be required to attain the desired effect.

Postsurgical strictures may arise at either the anastomotic site or in the adjacent oesophagus (Fig. 9.4). Their location and likelihood of occurrence is related to both the particular surgical procedure performed and its indication. Strictures of this type may be dilated whether they arise in the early postoperative period or at a later time as a result of scarring in association with inflammatory changes. Strictures that follow endoscopic sclerosis of varices or those resulting from prolonged nasogastric or tracheal intubation are also amenable to dilation. Strictures may develop following oesophageal perforation either from foreign bodies or iatrogenic causes (Fig. 9.5).

Strictures from mediastinal lesions or those which arise following radiation therapy directed at the head and neck are often difficult to negotiate without the advantage of radiographic guidance, and bougienage can be difficult, if not impossible. Malignancies involving the oesophagus may also be treated using radiographically assisted dilation, not for cure, but rather as a temporizing measure or to provide assistance in the placement of a feeding tube (Fig. 9.6). Even temporary relief of symptoms may produce a gratifying improvement in the quality of life of patients with these unfortunate conditions.

Review of surgical and endoscopic methods

Whether performed in the treatment of benign or malignant processes, procedures designed to restore or enhance oesophageal patency are generally classified as either surgical or nonsurgical. The following brief discussion of surgical and nonsurgical endoscopic treatment modalities is intended to highlight the fundamentals of these techniques in terms of their relative indications and limitations.

Surgical methods for managing oesophageal strictures, such as an antireflux operation in a patient with reflux oesophagitis, are directed at either the actual stricture or its underlying aetiology. Antireflux procedures are often very effective in ameliorating gastro-oesophageal reflux. However, surgical treatments which target the stricture primarily are often less satisfactory. Such operations generally entail either patching the oesophagus in the region of the stricture or bypassing the stricture with an interpositioned segment of colon or jejunum. While patching procedures may result in excellent relief of dysphagia and reflux symptoms, approximately half of those patients go on to require some type of dilation at a later date. Interposition operations are very effective, however, these major surgical procedures are associated with a recognized morbidity and mortality.

Bougienage, the passage of instruments through the oesophagus to effect subsequent stricture dilation, has been performed for centuries. In all forms, the technique entails passage of an instrument having a straight, tapered or olive-shaped tip through the stricture and the progressive stretching of the narrowed region until the desired lumen size is achieved. This technique can be performed either by the patient on multiple occasions using such bougies as the Hearst or Maloney dilators, or by a physician, with the use of an endoscope. The endoscopic approach led to the subsequent development of bougies which could be passed over guidewires, a technique which carried a smaller risk of perforation. Among the most popular of these in use at present are the Celestine and Eder-Puestow devices, both of which have a central lumen which allows for insertion over a previously positioned guidewire.

Advantages of the radiological techniques

While the aforementioned nonradiologically assisted methods effectively increase the size of the oesophageal lumen, they have several limitations. Surgical procedures carry initial success rates of up to 70%, but many patients require subsequent dilation to treat repeat stenoses. In addition, they necessitate general anaesthesia and carry a mortality rate which has been reported to be as high as 16% (Benedict, 1966).

Bougienage is associated with excellent results when used to treat technically straightforward lesions, but it carries a reported mortality rate of 1.5% and a complication rate of up to 8% (Westdorp *et al.*, 1982; Starck *et al.*, 1984; Lindor *et al.*, 1985). Traditional dilators effect an increase in lumen size by applying a stretching force to the oesophagus. However, they also exert a simultaneous longitudinal stress which has been implicated in mucosal tears and actual perforations. The use of angioplasty balloons (Benjamin *et al.*, 1982), in which a large radial force can be exerted on the lumen while keeping the longitudinal force minimal, has a distinct

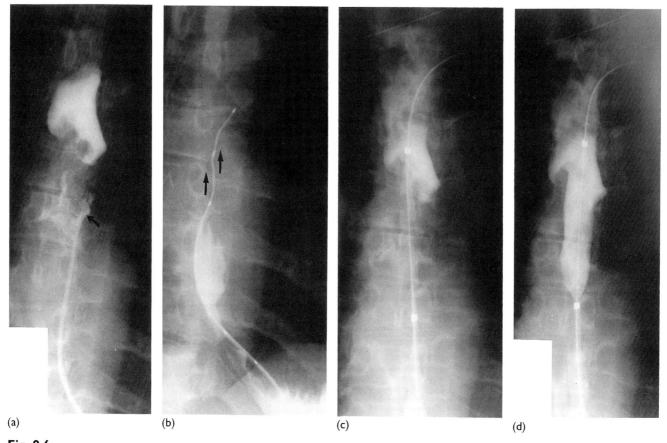

Fig. 9.6
A patient with oesophageal carcinoma, who had a gastrostomy in place. (a) In order to treat a tracheo-oesophageal fistula, dilation and placement of a Celestine tube was undertaken. A catheter has been placed through the gastrostomy and directed up into the tumour. The arrow indicates the position of the catheter tip. (b) A guidewire is torqued upward through the lesion. (c) A balloon catheter is placed over the guidewire. (d) The balloon is inflated. A Celestine tube was then positioned to stent the lesion.

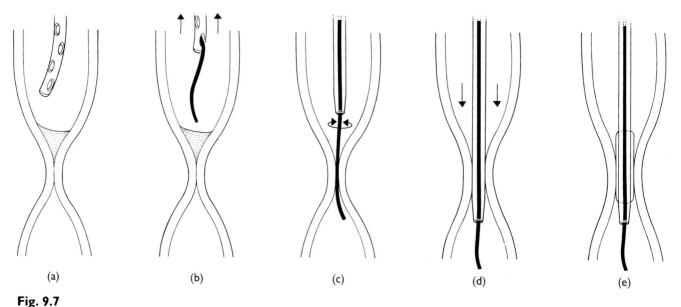

Fig. 9.7
Sequence illustrating the basic technique of oesophageal dilation. (a) A nasogastric tube is positioned proximal to the lesion. (b) A guidewire is passed through a side hole and the nasogastric tube is removed leaving the guidewire in place. (c) A catheter is inserted over the guidewire which is manoeuvred through the stricture. (d) The catheter is then advanced through the stricture, is removed and replaced with the balloon catheter. (e) The balloon is positioned at the region of the stricture and then inflated to the desired diameter.

advantage in many circumstances (McLean and LeVeen, 1989). Placement of traditional dilators, whether by patient or physician, can be technically difficult especially in cases involving a tortuous oesophagus associated with scarring or prior surgery. The sharply angled contours common to anastomotic strictures are difficult, if not impossible to negotiate with standard bougies, particularly if they are advanced without visual assistance.

The use of fluoroscopy, during the pre- and postprocedure evaluation and to monitor balloon catheter position, enables the radiologist to gauge the morphology and length of stricture and to identify immediately any complication. The latter point represents an obvious advantage, as early recognition of a complication permits greater latitude in selecting appropriate treatment.

CONTRAINDICATIONS

Contraindications to radiologically assisted balloon dilation of oesophageal strictures are no different from those of bougienage. Patients who are combative, or unable to cooperate are not candidates as patient input is important during the procedure.

Dilation at the site of an oesophageal perforation should be approached with caution. However, in the case of a proximal perforation associated with a distal stricture, dilation of the stricture may in fact facilitate healing by allowing secretions to pass beyond the perforation thus minimizing perioesophageal leakage.

While oesophageal malignancies are associated with an increased risk of perforation due to mucosal fragility, many may be dilated if care is taken to avoid overdilation. In some cases, conservative (i.e. less than 10 mm) dilation may be performed to allow insertion of a feeding tube which may then be used to stent the region of obstruction.

INSTRUMENTATION

The equipment required for balloon dilation is similar to that used in any vascular or biliary procedure. The typical armamentarium includes a variety of guidewires which aid in both traversing the stricture and in exchanging catheters through the stricture after it is crossed.

Catheters should be of standard configuration and may range from 4 to 7 Fr in diameter, depending upon the anatomy of the stricture and preference of the physician. Generally speaking, a straight catheter or a multipurpose 'hockey stick' configuration, is most useful in traversing oesophageal strictures. The basic technique is to use the catheter to support the wire, and to rely on the guidewire's configuration to permit passage through the stricture. Large bore catheters have the advantage of increased rigidity which may be helpful in situations where the anatomy is particularly distorted and tortuous, such as with anastomotic lesions.

Selection of the balloon catheter is dependent upon the stricture's morphology and aetiology. Generally speaking, low-pressure balloons are effective unless the stricture is sufficiently fibrotic to cause the balloon to rupture. We have not found that high-pressure balloons offer any distinct advantage otherwise. It is important to note that rupturing of these hydrostatic balloons is not associated with any increase in procedural morbidity.

Balloon catheters are available in a wide variety of sizes and the basic principles of selecting them will be discussed in the technique section. Generally speaking, with an extremely tight stricture, a 6 to 8 mm balloon, on either a 5 or 7 Fr catheter would be the initial instrument of choice. Increasingly large balloons will be selected as discussed later. The larger balloon sizes, for example, 12 to 20 mm, are all of the low-pressure variety and are available on larger shafts.

Selection of balloon length depends to a large extent on the morphology of the stricture, although in most cases the shorter balloons should be avoided unless required by anatomic restrictions such as a cervical oesophageal lesion. The shorter balloons tend to slide back and forth, which makes correct placement difficult. Those at least 8 to 10 cm in length are preferable. It is important to not overdilate any adjacent structures, as injury may result. However, the normal oesophagus can easily accommodate a 20 mm balloon without difficulty (Dawson et al., 1984).

When dilating the oesophagus of a patient with achalasia, some authors feel it is necessary to overdistend the stricture, which may require the use of multiple balloons (Starck et al., 1984). The use of two balloons will result in an oblong-shaped dilation surface (Gaylord et al., 1988), while three balloons in tandem present a more rounded surface which more closely approximates normal oesophageal morphology. However, in our experience, repeated dilation to 20 mm is generally adequate.

Technique

The following discussion outlines and updates the fundamental techniques of oesophageal dilation (Meranze and McLean, 1987; Cope *et al.*, 1990). Each case presents its own unique set of circumstances which the interventional radiologist must approach in a logical manner.

Regardless of the dilation technique employed, it is essential first to obtain a complete history to determine both the aetiology of the stricture as well as whether placement of a feeding tube will be required. A complete oesophagogram in multiple projections should be performed to outline the morphology of the stricture. This also permits identification of any associated abnormalities such as dysmotility, diverticula or reflux. In many cases endoscopy is performed prior to dilation in order to obtain biopsies of the abnormal region, although it is also possible to biopsy a stricture using biopsy brushes during the fluoroscopic phase of a balloon dilation.

We recommend placing an intravenous line in all patients to permit administration of medications as required. In most cases only mild sedation, such as intravenous diazepam (Valium) or midazolam (Versed), is required. It is important not to oversedate the patient, not only to avoid the possibility of aspiration but also because a conscious patient provides valuable assistance when making a judgement as to how aggressively the stricture should be dilated. While it is usual for patients to experience some degree of discomfort, the occurrence of severe pain should be an indication to discontinue the procedure to avoid the possibility of oesophageal rupture.

Prior to performing the initial intubation, we generally spray the oropharynx with a topical anaesthetic. A soft-tipped nasogastric tube is placed either transorally or transnasally to a level just above the stricture. This tube will need to be modified prior to placement to allow an exchange guidewire to be passed through it. By enlarging a side hole in the nasogastric tube, rather than opening the end, its rounded tip can be preserved and permit the insertion to be less traumatic.

The decision between use of the transnasal or transoral route depends largely on whether a soft-tipped nasoenteric feeding tube will be required after the dilation. If placement of such a feeding tube is indicated, the transnasal approach is preferred. If feeding tube placement is not a consideration, we generally prefer the transoral approach as it is more technically straightforward, particularly when using larger ballons and, by and large, is better tolerated by the patient. With either approach, the insertion of the nasogastric tube is performed in a standard fashion and is often expedited by having the patient flex their neck and swallow small mouthfuls of water.

Once the nasogastric tube is in position, a small amount of water-soluble contrast material is injected through the tube to delineate the proximal aspect of the stricture. In the case of a patient at increased risk for aspiration, either dilute barium or an oily contrast medium can be used to prevent pulmonary complications. After the proximal portion of the stricture has been outlined, a guidewire of either a torquable type or exchange type, is passed through the tube exiting the enlarged side hole. The tube is removed and an angiographic catheter is inserted over the guidewire. As mentioned previously, we generally use a straight or multipurpose-type catheter, although any catheter with a mild bend may be used. Once the catheter is directed towards the lumen, a flexible guidewire generally passes easily through the stricture. If the lumen is irregular, sometimes a small J curve, such as a Rosen guidewire can be used to traverse the stricture.

In difficult cases, torquable guidewires might be preferred. The Glide wire (Terumo) or the Wholey wire (ACS) are both examples of torquable guidewires that can be used to traverse strictures. It should be noted that if the Glide wire is chosen, one needs to exert particular caution to not advance it too aggressively when there is any possibility of submucosal passage, as the slipperiness of this device makes it relatively easy inadvertently to perforate the oesophageal wall. Patients with malignancies are at particular risk for perforation and frequent injections of contrast material should be made during the catheter passage phase to ensure appropriate position.

After the guidewire has advanced through the stricture, the catheter can be directed over it and placed so that its tip is beyond the distal aspect of the stricture. A stiffer exchange length guidewire can be inserted through the catheter, and the catheter removed. The balloon catheter is then passed over this guidewire such that the balloon is centred in the stricture.

The selection of a particular balloon catheter depends largely on the stricture morphology as determined by oesophagoscopy and, more importantly, on the findings of prior barium studies. The final goal should be a 20 mm lumen, although the initial size of the stricture will be a large factor in determining the first balloon chosen. If an extremely tight lesion (i.e. less than 3 mm) is present, we will generally start with a 6 to 8 mm balloon. If there is no 'waist' present during the dilation, and if the patient experiences no discomfort, we will increase the balloon size. Once the balloon is inflated we will generally maintain it in the inflated state for approximately 1 min with the use of a stopcock device or LeVeen inflator (Medi Tech, Inc.). A prolonged inflation is particularly important in fibrotic areas having substantial scarring as, in contrast to vascular dilation procedures, the 'waist' will often not disappear rapidly but may, instead, disappear over a period of time.

In addition to achieving the desired goal of a 20 mm lumen, the patient's response to the dilation procedure is one of the most important factors in determining the maximal balloon size. If the patient experiences sharp pain, the procedure is generally terminated and a dilute barium or Gastrografin swallow is performed to evaluate the possibility of perforation. No further dilations should be performed at that time, but most patients can safely return for further dilations in several days time.

Following the procedure, a complete oesophagram is performed using either dilute barium or Gastrografin. This is intended more to rule out the possibility of perforation or mucosal tear than to assess the success of dilation. The luminal appearance immediately following dilation may suggest a poor result when, in fact, the patient will go on to report symptomatic relief after some of the postdilation oedema subsides.

In general, the number of dilations required for symptomatic relief varies with lesion aetiology and morphology. In one series (Stark et al., 1984), 58% of patients required only one session, with another 19% requiring a second treatment. As with other forms of dilation, such as bougienage, the procedure may need to be repeated many times over a period of time. Thus, it is most important to approach the dilation with realistic expectations and the understanding that the need for repeat procedures is not considered a failure. Repeat dilations can be performed after several days or weeks, depending on the severity of both the patient's symptoms and the lesion at the time of dilation.

If the patient presents with a gastrostomy tube in place, it is often feasible to use this as a means of access, particularly if it is directed towards the cardia. In such cases we have used a vascular-type access sheath, through which catheters and balloons are inserted. The indwelling gastrostomy tube is removed and the sheath substituted for it. This allows easy maintenance of access during catheter exchanges. It is important, however, to make certain that the gastrostomy is mature and it is therefore prudent to not attempt such exchanges unless the gastrostomy has been in place for at least several weeks.

Balloon dilation can be of great benefit to those with anastomotic strictures as they are often not ideal surgical candidates and conventional endoscopic techniques are likely to be difficult. A complete review of all X-rays as well as any operative reports is extremely helpful in the pretreatment planning of these patients.

The basic techniques are similar to those outlined previously, with the particular challenge of anastomoses being associated more with the increased difficulties of intubation than with the dilating procedure itself (McLean et al., 1986). Larger, thicker walled catheters of 6 F or greater, are stiffer and are often easier to control in a large lumen such as that present following a colonic interposition. If necessary, endoscopic assistance can be employed to treat these strictures. The endoscope assists in locating the lumen and in supporting the passage of the balloon catheter. Balloons of 12 to 15 mm in size are generally adequate for most anastomoses.

COMMON MISTAKES AND HOW TO AVOID THEM

While the techniques described above are fairly straightforward, a few points can help to make the procedures as trouble free as possible.

The initial intubation through a transnasal approach is often uncomfortable for the patient. The use of an anaesthetic gel can go a long way in avoiding discomfort. We find that by directing the tube horizontally, or even slightly downward through the nose, this initial portion of the intubation is simplified. If a transoral approach is used, the liberal use of a topical anaesthetic spray will aid in preventing gagging.

When choosing a balloon, a common problem is selecting too short a balloon, which can allow a 'to and fro' motion. An 8 cm length is a good size for most situations, although in very proximal lesions, 4 cm may be all that is technically feasible. If there is sharp angulation in the region of the stricture, it may be necessary to use a shorter balloon. In either case, by advancing the balloon slightly off centre distally, and then applying traction to the catheter during inflation, it is often possible to keep the balloon centred during dilation.

Although we have not experienced complications while dilating strictures from caustic ingestion, it is important to proceed with caution in these patients. By allowing at least several weeks to elapse from the time of ingestion before attempting dilation, one can be more confident that the initial inflammatory response will have abated, and some scarring will have occurred. This should increase the margin of safety.

When dealing with anastomotic lesions, it is often difficult to differentiate between a functional impairment and an anatomic stricture. Frequently, there is abnormal motility due to the recent surgery, and it is therefore hard to evaluate the region of interest adequately. In those cases, a trial dilation may be justified, as there is little risk and it is possible to unmask a hidden lesion.

It is important to remember that the post-procedure films should not be used to assess the degree of success following dilation. Rather, they should only serve as a screen for significant mucosal lacerations. The patient should decide whether additional therapy is indicated based on the degree of symptomatic relief.

COMPLICATIONS

Complications are extremely uncommon. Minor complications would include vasovagal episodes due to discomfort which can be prevented by using appropriate sedatives and, if necessary, analgesics. In addition, a patient with a history of vasovagal reactions may benefit from the administration of intravenous atropine prior to the procedure. As was mentioned previously, we always start an intravenous line with normal saline solution prior to the procedure to enable administration of any needed medications.

Aspiration of Gastrografin can occur if the lesion is particularly high grade or if the patient experiences intense gagging. If aspiration is suspected, use of dilute barium would be preferred over Gastrograffin for any swallows.

The most serious complication would be perforation of the oesophagus either with a guidewire while traversing the lesion, or as a result of the dilation itself. Most small perforations will heal without further treatment. In one series, there were three perforations, two of which occurred in malignant strictures (Gaylord *et al.*, 1988) with one of these requiring surgery. In another series, two oesophageal tears were reported following anastomotic stricture dilation. Both resolved with conservative management. If a large perforation should occur, supportive measures including antibiotics and

fluids should be administered while surgical consultation is obtained. In cases of perforation following endoscopic dilation, we have inserted drainage tubes transnasally and through the oesophageal tear into the collection. However, this complication should be extremely rare if the basic guidelines outlined above are followed and close attention is paid to patient discomfort during the procedure. For this reason, uncooperative patients should be approached very cautiously and balloon size should be selected accordingly. Minor amounts of blood on the balloon following the procedure should not be a cause for alarm, although profuse bleeding is uncommon and should be investigated.

There has been some discussion concerning the need for radiation (fluoroscopy) in the performance of these studies. With modern image intensification equipment, a skilled radiologist can perform all but the most difficult cases with minimal patient exposure. The added control and margin of safety afforded by fluoroscopic guidance greatly outweighs any potential harm from the radiation.

POSTPROCEDURAL CARE

Postprocedure we place few restrictions on the patient. We allow them to increase their diet as tolerated unless there is evidence of a mucosal tear. In that case, we would have them withhold all oral intake for 12 hours and then increase to liquids only for another 24 hours. At that point, a repeat oesophagogram is performed to evaluate healing. In most cases, unless there is malignancy, this would be adequate time for repair.

We follow-up all patients over the few weeks following dilation to evaluate the symptomatic success as well as to monitor for the occurrence of delayed complications. If it is clear that additional treatment sessions will be required, these are scheduled at the time of follow-up.

Results

The efficacy of balloon dilation has been examined in a number of series. Published success rates range from 67% to close to 90%, depending on the type of lesion and duration of follow-up. Complications have been few, with one centre reporting two lacerations, both of which healed with conservative treatment (de Lange *et al.*, 1987).

McLean *et al.* (1987) evaluated the results of 94 dilations (50 of which were oesophageal) over a 6-year period with an average duration of follow-up in excess of 1 year. Data were analysed to assess the impact of procedural factors on outcome. Intubation was technically successful in 98% of oesophageal lesions. Of note, final symptomatic outcome was not correlated with a variety of procedural aspects including ease of intubation, postdilation appearance, and the number of dilations required. On average, oesophageal lesions required 2.2 treatment sessions for a satisfactory result. Eighty-one per cent of patients were symptom-free at 6 months, although this decreased to 71% at 9 months. Others have reported up to 92% of patients being symptom-free at 12 months.

Conclusions

While traditional techniques are often effective in the treatment of oesophageal strictures, radiographically guided therapy can offer distinct advantages in many situations. By becoming skilled in these straightforward and effective procedures, the interventional radiologist can provide an important service to both the patient and his physician confronted with this difficult problem.

References

BENEDICT EB (1966). Peptic stenosis of the esophagus: A study of 233 patients treated with bougienage, surgery, or both. *American Journal of Digestive Disease* **11**: 761–70.

BENJAMIN SB, CATTAU EL, GLASS RL (1982). Balloon dilatation of the pylorus: Therapy for gastric outlet obstruction. *Gastrointestinal Endoscopy* **28**: 253–4.

COPE C, BURKE DR, MERANZE S (1990). *Atlas of Interventional Radiology*, pp. 12.1–12.17. J.B. Lippincott Company, Philadelphia.

DAWSON SL, MUELLER PR, FERRUCCI JT JR, RICHTER JM, SCHAPIRO RH, BUTCH RJ, SIMEONE JF (1984). Severe esophageal strictures: indications for balloon catheter dilation. *Radiology* **153**: 631–5.

DE LANG EE, SHAFFER HA, DANIEL TM, KRON IL (1987). Esophageal anastomotic leaks: Preliminary results of treatment with balloon dilation. *Radiology* **165**: 45–7.

GAYLORD GM, PRITCHARD WF, CHUANG VP, CASARELLA WJ, SPRAWLS P (1988). The geometry of triple-balloon dilation. *Radiology* **166**: 541–55.

LINDOR KD, OTT BJ, HUGHES RW JR (1985). Balloon dilatation of upper digestive tract strictures. *Gastroenterology* **89**: 545–8.

LONDON RL, TROTMAN BW, DIMARINO AJ JR, OLEAGA JUAN A, FREIMAN DB, RING EJ, ROSATO EF (1981). Dilatation of severe esophageal strictures by an inflatable balloon catheter. *Gastroenterology* **80**: 173–5.

MCLEAN GK, COOPER GS, HARTZ WH, BURKE DR, MERANZE SG (1987). Radiologically guided balloon dilation of gastrointestinal strictures. *Radiology* **165**: 35–43.

MCLEAN GK, LEVEEN RF (1989). Shear stress in the performance of esophageal dilation: Comparison of balloon dilation and bougienage. *Radiology* **172**: 983–6.

MCLEAN GK, MERANZE SG, BURKE DR (1986). Enteric alimentation: A radiologic approach. *Radiology* **160**: 555–6.

MERANZE SG, MCLEAN GK (1987). Interventional techniques on the alimentary tube. In: *Radiology: Diagnosis, Imaging, Intervention*, pp. 1–9. Edited by Taveras JM, Ferrucci JT. J.B. Lippincott Company, Philadelphia, Volume 4, Chapter 9.

STARCK E, PAOLUCCI V, HERZER M, CRUMMY A (1984). Esophageal stenosis: treatment with balloon catheters. *Radiology* **153**: 637–40.

WEBB WA, MCDANIEL L, JONES L (1984). Endoscopic evaluation of dysphagia in two hundred and ninety-three patients with benign disease. *Surgery, Gynecology and Obstetrics* **158**: 152–6.

WESDORP JC, BARTELSMAN JW, DEN HARTOG JAGER FCA, HUIBREGTSE K, TYTGAT GM (1982). Results of conservative treatment of benign esophageal strictures: A follow-up study in 100 patients. *Gastroenterology* **82**: 487–93.

CHAPTER 10

Gastrointestinal strictures and fistulae

Dana Burke and Gordon K. McLean

Fluoroscopically guided balloon dilation of gastrointestinal strictures   141

Indications   141

Patient preparation   141

Equipment   142

Technique   142

Results   154

Complications   158

Percutaneous management of gastrointestinal fistulae   158

Diagnostic studies   159

Conservative management   160

Percutaneous intervention   160

Results   165

Reference   170

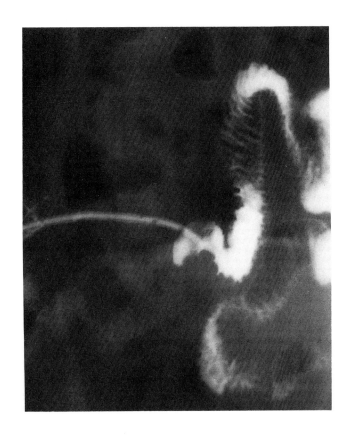

FLUOROSCOPICALLY GUIDED BALLOON DILATION OF GASTROINTESTINAL STRICTURES

Traditionally, gastrointestinal strictures have been treated with surgical resection or bypass. Although surgical treatment is generally associated with a high success rate, there is also the potential for a significant morbidity and mortality, particularly in debilitated patients with multisystem disease (Herrington and Sawyer, 1990).

Radiological management of enteric strictures was first described by London *et al.* (1981) in an article which described their success with fluoroscopically guided balloon dilation of oesophageal strictures (London *et al.*, 1981). Since that time, a body of literature has accumulated which documents the ease of fluoroscopically guided balloon dilation at many different sites along the gastrointestinal track (Starck *et al.*, 1984; de Lange and Shaffer, 1988, 1991). The superiority of balloon dilation over co-axial dilation is evidenced by the fact that this technique has largely supplanted the use of tapered dilators in many gastroenterological practices. Tapered dilators produce considerable shear force at dilatation sites, pose an increased risk of perforation and are relatively inflexible, limiting their application to lesions in the oesophagus (McLean and Le-Veen, 1989). In addition, although the endoscope demonstrates the proximal end of a stricture quite well, the way through the stricture may not always be apparent. By using iodinated contrast medium and image-intensified fluoroscopy, the radiological method has the advantage of displaying the entire course and length of the stricture (Fig. 10.1).

Because of the limitations and risks of both surgical and endoscopic management, fluoroscopically guided techniques have been developed over the past 10 years allowing rapid intubation followed by cannulation and balloon dilation of gastroenteric strictures at multiple sites. Fluoroscopically guided balloon dilation combines a high success rate with minimal morbidity and is the least expensive and least invasive approach to the treatment of gastrointestinal strictures.

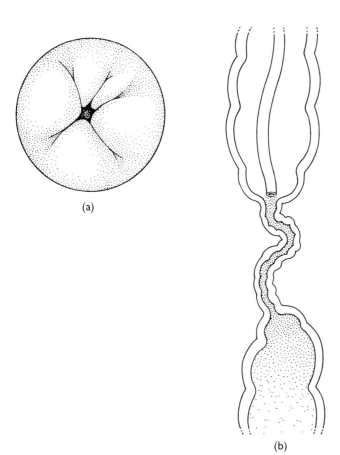

Fig. 10.1 *Differences in endoscopic and fluoroscopic perspective*
(a) The endoscopic approach clearly demonstrates the origin of the stricture, but provides little information regarding what lies distal to the orifice. (b) Contrast medium injection through a catheter at the orifice of the stricture clearly delineates the length, course and calibre of the strictured segment. This additional information contributes to safety of fluoroscopically guided cannulation of their regularly strictured segments.

INDICATIONS

Any stricture which can be reached with a balloon catheter can be treated with balloon dilatation. As there is no well-defined group of patients that predictably fail to benefit from fluoroscopically guided balloon dilatation interventional radiologic treatment of any stricture which can be reached should be attempted before more invasive procedures such as endoscopy or surgery are undertaken.

PATIENT PREPARATION

Patients should have nothing by mouth for at least 4 and preferably 8 hours before the procedure. Mild sedation is occasionally required for anxious patients, and is best provided by short-acting agents such as midazolam (Versed Roche Laboratories, New Jersey) 0.5–1 mg, administered intravenously. Preprocedure broad-spectrum antibiotics are recommended, particularly for those undergoing rectal or colonic dilations.

EQUIPMENT

Fluoroscopically guided cannulation and dilatation can be performed readily in any standard fluoroscopic room with a good quality image intensifier and filming capability. A tilting table and/or C-arm frequently facilitate performance of the procedure, but are not absolute requirements. The guidewires and catheters used in these procedures are generally available in most angiography suites, although extra-long guidewires of 200 to 300 cm length may be required. Commonly used equipment is listed in Table 10.1.

TECHNIQUE

Gastrointestinal access

GASTRIC INTUBATION

The procedure begins with anaesthetization of the oropharynx. The patient is asked to protrude his tongue which is then grasped by the examiner with a dry gauze while a topical spray anaesthetic (Cetacaine, Cetylite Industries, Inc., Pennsauken, NJ) is applied. The tongue is then released and the patient is asked to gargle briefly with the residual anaesthetic before swallowing. This produces thorough anaesthetization of the oropharynx and proximal oesophagus.

A 10 Fr nasogastric tube is then modified by cutting a large hole several centimetres proximal to the tip with a pair of scissors. The tip of this tube is then coated with viscous Lidocaine (Roxane, Inc., Columbus, Ohio). The tube is advanced through the mouth and down the oesophagus while the patient is instructed to swallow to facilitate its passage. Once the nasogastric tube has reached the stomach, as confirmed fluoroscopically, an exchange length guidewire with a soft curved tip is advanced through the tube lumen and out of the enlarged side hole

Table 10.1 *Basic equipment for balloon dilatation of gastrointestinal strictures*

4 × 4 Gauze
Cetacaine Spray
10 French nasogastric tube
Viscous Lidocaine
Exchange length (200–300 mm) guidewires
100–125 cm length torqueable catheters
Torqueable guidewires
   Lunderquist-Ring
   Glide wire
   Wholey wire
Enteroclysis tube
   Herlinger tube
   Bilbao – Dotter tube
Teflon dilators 8–20 French
Lead Glove
Balloon catheters 8–20 mm.

followed by removal of the nasogastric tube. A 100 cm long torqueable catheter is then advanced over the guidewire for subsequent manipulation through the lumen of the stomach to the gastric outlet. The catheter is directed towards the pylorus or gastroenterostomy and the very soft-tipped guidewire is passed through it. The guidewire will tend to buckle when it contacts the wall of the stomach and with the appropriate orientation of the catheter tip it can be steered in the desired direction. The catheter is advanced over the wire to the point closest to the ultimate destination and the wire is then withdrawn back into the catheter, the catheter reoriented and the wire advanced again. This process is repeated until the catheter reaches the gastric outlet (Fig. 10.2).

GASTROSTOMY INTUBATION

In some patients, a pre-existing gastrostomy tube will provide alternative access to the gastrointestinal tract. It is necessary to remove the gastrostomy tube before proceeding with the cannulation of the pylorus. However, not all surgical or endoscopically placed gastrostomy tubes are removable percutaneously due to a fixed, intraluminal locking mechanism and this should be ascertained before choosing this access route. The gastrostomy tube area is cleaned and covered with sterile towels. A stiff guidewire such as a Lunderquist–Ring torque guidewire (Cook, Inc., Bloomington, IN) is then passed through the tube. After removal of the

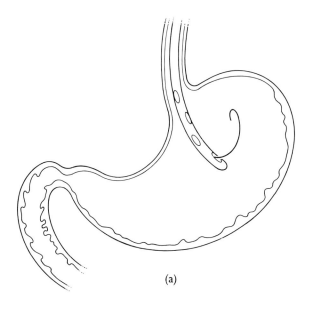

(a)

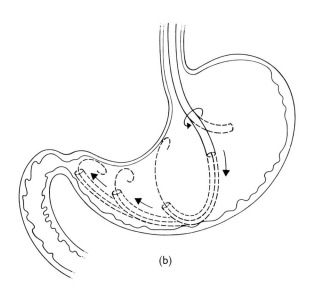

(b)

Fig. 10.2 *Gastric intubation and advancement to the pylorus*
(a) Following intubation of the stomach, a curved guidewire is
manoeuvred out of the enlarged side hole into the lumen of the
stomach. The feeding tube is removed and replaced with a
torquable catheter. (b) The catheter is turned to the orientation
which leads away from the fundus and towards the body of the
stomach. The soft-tipped guidewire is then advanced and buckles
within the stomach lumen, with the buckling toward the pyrlorus.
The catheter is advanced along the wire to the point closest to the
pylorus, the wire is withdrawn into the catheter, the catheter
reoriented and the guidewire again advanced to the point along the
wire closest to the pylorus. This sequence of catheter-directed
guidewire buckling followed by catheter advancement is repeated
until the pylorus is reached.

towards the fundus rather than the pylorus. Manipulation through a catheter that is looped in the fundus
is limited as is the ability to apply forward pressure.
Therefore, the challenge is to redirect the catheter
and guidewire directly from the gastrostomy site to
the pylorus. This is done with the torquable catheter
in conjunction with soft-tipped guidewires as previously described. In some cases, the angles between
the gastrostomy tract and the body of the stomach
are so steep that it is not possible to reorient with the
guidewire and catheter. In these cases, a metal stiffening cannula or rigid Teflon dilator can be used to
redirect the guidewire and/or catheter toward the
pylorus. Once the catheter and guidewire have
reached the pylorus, exchange is made for a stiff
Amplatz guidewire (Cook Inc., Bloomington, IN)
and a 12 to 15 Fr sheath is placed which extends from
the gastrostomy site directly to the pylorus. This
effectively reduces the space in which catheter and
wire manipulation is performed from the large
volume of the stomach to the small tubular space of
the sheath lumen. This technique prevents buckling
of the catheter and wire into the fundus with further
manipulation, an occurrence which frequently
results in inadvertent retraction of the wire and
catheter tip, and always results in decrease in torque
control (Fig. 10.3).

TRAVERSING THE PYLORUS

Whether the pylorus is strictured or not, it will
frequently present some difficulty in cannulation.
The combination of catheter curve and guidewire
characteristics that will succeed in traversing the
pylorus varies greatly among patients due to
anatomical variations of the gastric antrum, pylorus
and duodenum. Often a curved angiographic catheter such as the cobra, multipurpose or Berenstein
shapes will succeed in conjunction with a soft-tipped
guidewire. The catheter is simply torqued in a
variety of directions followed by advancement of the
wire. In nonstrictured pyloric channels, the soft-
tipped guidewire will usually traverse the pylorus
when the correct orientation of the catheter has been
chosen. If this method is unsuccessful because the
wire repeatedly coils in the antrum or buckles retrogradely, the soft-tipped wire is replaced with a
stiffer, torquable wire (e.g. Lunderquist–Ring,
Cook Inc., Bloomington, IN or stiff Glide Wire,
Terumo, Japan). These wires are actively steered
through the irregular or angulated pyloric channels,
rather than buckled, and will often succeed when
softer tipped wires have failed. In patients with large
dilated stomachs or with pyloric anatomy distorted
by inflammatory disease or surgery, larger catheters

gastrostomy tube, a torquable catheter is inserted. In
most surgically and endoscopically placed gastrostomy tubes, the orientation of the tube tract is

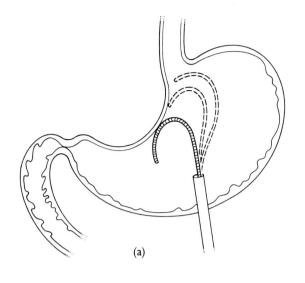

(a)

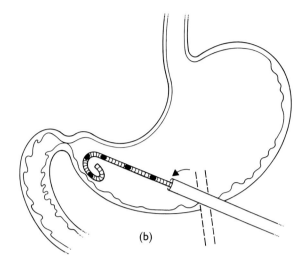

(b)

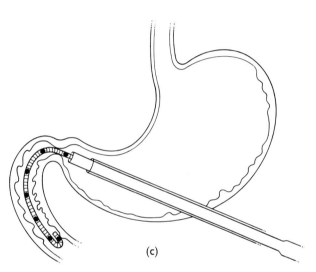

(c)

Fig. 10.3 *Use of gastrostomy tube site as access for upper gastrointestinal balloon dilation*
(a) After the gastrostomy tube is removed over the guidewire, the course of the gastrostomy tract usually orients the guidewire and catheter towards the fundus. Although looping the catheter and wire through the fundus and then out the pylorus can be done, this limits torquability. (b) If it is not possible to advance a standard catheter and wire to the pylorus without buckling, the catheter is replaced with a rigid metal cannula or Teflon dilator. By levering the outside portion of the cannula or dilator, the transgastric track can be reoriented toward the fundus. A Lunderquist–Ring guidewire is then passed through towards the pylorus. (c) Over a stiff guidewire the transgastric track is dilated and a 12 to 15 Fr sheath put in place which extends from the gastrotomy site to the pylorus. Subsequent catheter and guidewire manipulations are performed through the sheath to prevent buckling into the fundus.

designed for small bowel intubation such as the Bilbao–Dotter or Herlinger enteroclysis tubes (Cook Inc., Bloomington, IN) may be very useful. These are flexible large bore tubes through which heavy torque cables are passed which provide both the steerability and rigidity that facilitates traversing the pylorus while resisting buckling within the stomach. At times, the exertion of gentle pressure on the stomach with a lead glove to minimize stretching of the stomach or buckling of the catheter can be quite useful (Fig. 10.4).

ADVANCING THROUGH BOWEL

Once the pylorus or gastrojejunostomy has been traversed, progress is made through the lumen of the small bowel using gently curved catheters and soft-tipped guidewires. These are much more useful than sharply curved catheters such as the cobra catheter or relatively stiff-tipped torquable guidewires which will tend to catch on the valvulae conniventes and make advancement very difficult. Excessive pressure on the stiffer wires in this situation can lead to buckling of the catheter within the lumen of the stomach. When using a stiff wire if a loop forms within the stomach, this will result in sudden retraction of the tip of the catheter back through the pylorus or gastrojejunostomy, losing considerable ground. Unlike its stiffer counterpart, the wire proximal to the tip buckles forward with additional pressure rather than causing catheter buckling in the stomach. The redundant wire presents a smooth, blunt surface which is no longer retarded by the valvulae as it is advanced (Fig. 10.5).

As in the stomach, the guidewire is advanced to the point where additional pushing causes retraction

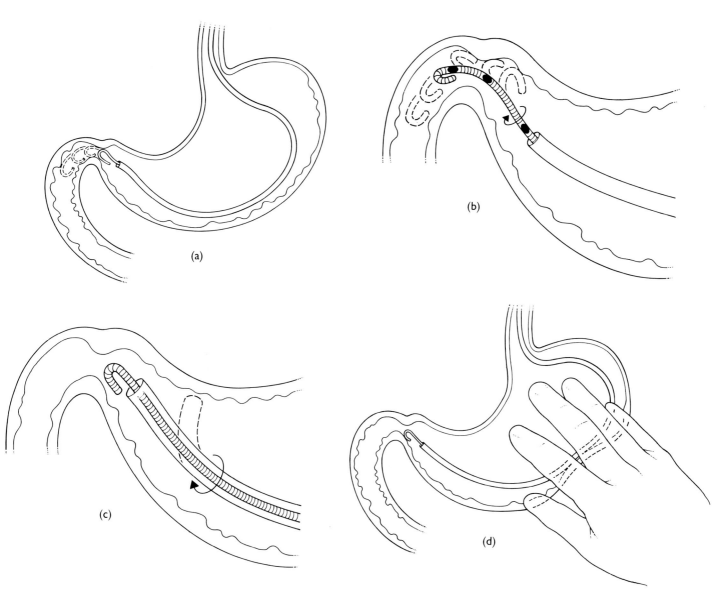

Fig. 10.4 *Crossing the pylorus*
(a) Frequently a soft-tipped guidewire will buckle through the nonstrictured pylorus when the correct orientation of the torquable catheter is achieved. (b) Narrow or irregular pyloric channels may require the combination of torquable catheter and guidewire. With the catheter oriented towards the pyloric channel, the torquable wire is steered through the pylorus and into the duodenum. (c) In very dilated or J-shaped stomachs, there can be considerable loss of torque and a tendency towards buckling with standard wires and catheters. The use of large (15 Fr) catheters with large calibre torque cables will provide both additional torque control and resistance to buckling and will usually allow successful cannulation when standard methods fail. (d) The lead-gloved hand can be used to facilitate pyloric cannulation with any technique as compression of the greater curvature will limit stretching of the stomach and improve the angle through which torque is transmitted. It also allows one to direct more pushing force to the tip of the catheter.

of the catheter tip. At this point, the guidewire is held steady and the catheter advanced to the leading portion of the curve. The cycle of buckling forward with the guidewire followed by advancement of the catheter is repeated until the stricture site is reached.

It is important to be patient when advancing catheters and wires through the bowel lumen as peristalsis and the distensibility of the bowel wall will frequently result in delayed response to attempts at forward movement. Once the stretched gut wall recoils or the peristaltic wave which temporarily pins the wire and catheter passes, the catheter and guidewire will advance spontaneously, often 10 to 20 sec after the operator last applied any forward pressure.

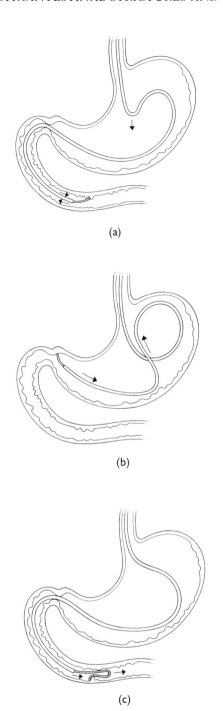

Fig. 10.5 *Advancing through bowel*
(a) Stiff-tipped guidewires will catch on valvulae conniventes as they are advanced through the small bowel or the haustra within the colon. Continued pushing on the stiff guidewire will result in buckling of the catheter proximally which causes retraction of the catheter tip. (b) If this buckling is not recognized, additional pushing will result in a loop of catheter forming suddenly within the stomachs and sudden retraction of the catheter and guidewire tip, often back into the stomach itself. (c) In contrast, a soft-tipped guidewire buckles forward if its tip catches in the valvulae and the blunt leading edge thus formed will travel unimpeded through the bowel. After the advancement of 10–20 cm of wire, the catheter is then advanced over the wire while the wire is held steadily or is slightly withdrawn.

RECTAL INTUBATION

Intubation of the colon is usually simpler than that of the upper gastrointestinal tract. The patient is placed in the left decubitus position with the right leg flexed at the hip and the knee. A lubricated red rubber catheter modified with a large side hole is introduced. A curved soft-tipped guidewire is advanced through the red rubber catheter and exchanged for a gently curved catheter. Because of the large calibre and redundancy of the colon, the use of positive contrast medium to delineate the anatomy should be kept to a minimum as accumulated contrast medium may obscure the way through a redundant colon. Air introduced through the catheter will usually help to define the anatomy without obscuring the way, particularly if a diagnostic barium study is available to provide a 'roadmap'.

The guidewire and catheter are used in conjunction utilizing the same technique as described for advancing along the small bowel. The cycle of guidewire advancement and catching up with the catheter is repeated until the site of stricture is reached. Because of the larger calibre of the colon, more directional help from the steerable catheter is often required to prevent coiling of the wire within a bend in the bowel lumen. Once the site of stricture is reached, Gastrografin or other water-soluble contrast agent may be injected to confirm its location as well as define its contours and length.

Cannulation and balloon dilation of strictures

GASTRIC OUTLET STRICTURES

The most commonly treated non-oesophageal strictures are those which involve the gastric outlet, either of the pylorus or of a surgical anastomosis. Pyloric and antral scarring is usually the result of long-standing peptic disease, but occasionally can result from ingestion of strongly acidic or caustic substances, or as a complication of pyloric surgery. Surgically created gastroenterostomies and surgical procedures for the treatment of morbid obesity that involve restriction of gastric volume and/or jejunal bypass are other settings in which gastric outlet obstruction can occur.

Pyloric strictures
It is reasonable to attempt briefly negotiation of pyloric and antral strictures with the combination of catheter and soft-tipped guidewire used to approach

the stricture, but frequently this will result in buckling of the soft wire back into the lumen of the stomach rather than successful cannulation. Torquable guidewires are usually required to cannulate pyloric strictures successfully. There are several available torquable wires with different characteristics. The Lunderquist–Ring wire (Cook Inc., Bloomington, IN) is stiff throughout its length, with solder points over the last several centimetres which allow precise shaping of simple or complex curves. The Wholey wire (Advanced Cardiovascular Systems, Temecula, California) has a relatively stiff shaft with gradual transition to a very floppy curved tip. The Glide wire (Terumo, Japan) has a unique construction of a solid metal core coated with polyurethane to which a hydrophilic coating is bonded. The specific wire chosen will depend on the operator's preference and previous experience. After the catheter tip has engaged the proximal portion of the strictured channel, a small amount of contrast medium is injected to delineate the course and length of the lumen to be cannulated. The torquable guidewire is then manipulated through the channel and into the duodenum. It is important to advance a sufficient length of wire beyond the stricture before attempting to advance the catheter. The appropriate length of wire for this purpose varies with wire design. The goal is to have sufficient stiff guidewire distal to the stricture to allow advancement of the catheter through the stricture. If only a soft guidewire is passed beyond the stricture, it will tend to buckle back into the stomach.

Once the catheter has been advanced through the stricture, exchange is made for a long (i.e. 210 to 260 cm) stiff guidewire such as the Amplatz exchange wire or Coons Interventional wire (Cook Inc., Bloomington, IN). These are very stiff-shafted, soft-tipped wires which will move forward through the bowel using previously described techniques, but have sufficient body to carry catheters through tight strictures without buckling. After the exchange guidewire is in position, the diagnostic catheter is replaced with the balloon dilating catheter. The catheters used for gastrointestinal dilation are similar in design to those used in angioplasty, but have longer balloons (8 cm) for increased stability during dilation. Shorter balloons tend to migrate proximally or distally during expansion. The chosen balloon is slowly expanded at low (3 atm) pressure to the end point of dilation and maintained at the diameter for 3 min. The goal of pyloric dilation is to achieve full expansion of a 15 mm balloon, as this will usually relieve obstructive symptoms (Fig. 10.6). Some pyloric strictures are very resistant and may require sequential dilation, starting with smaller balloon diameters and gradually reaching 15 mm in one or more sessions. The end point of dilation at a given session is determined by patient tolerance as the stretching involved with balloon dilation can be uncomfortable and significant discomfort is a reasonable end point to use. Severe pain is to be avoided as too rapid or too large a dilation can increase the risk of tearing the bowel wall. The total number of dilation sessions will be determined by the ability to reach targeted balloon size and the degree of symptomatic improvement achieved. The goal is to allow return to and maintenance of a regular diet.

Gastroenteric strictures

The most common cause of gastric outlet obstruction treated with balloon dilation is stricturing following surgical gastroenterostomy. Gastroenterostomies must be precisely constructed to allow adequate gastric emptying, but the size of the anastomoses must be small enough to prevent a dumping syndrome. Anastomotic obstruction occurs in about 5% of patients treated with gastroenterostomies (Herrington and Sawyers, 1990). Because of the many different surgical techniques involved, as well as differences in patient anatomy, gastroenterostomies are the most difficult strictures to cannulate. Preprocedure review of available imaging studies, particularly barium studies and computed tomographic (CT) scans, can be very useful in choosing the optimum patient orientation as well as directing the initial choice of guidewires and catheters. In general, gastroenterostomies are constructed on the anterior surface of the stomach, and a steep oblique projection will often best display the anastomotic channel. It is in this group that the C-arm can be most helpful for obtaining the ideal orientation of the patient in relation to the fluoroscopic beam. The methods used do not vary significantly from those used to cross the pylorus, but are technically more demanding. Torquable guidewires and/or large bore enteroclysis type tubes are routinely required. In contrast to the intact stomach where the calibre of the stomach narrows gradually toward the pylorus, the gastric remnant with a stenotic gastroenterostomy resembles a sphere with a pinhole, often in a completely nondependent location. Locating the tightly strictured anastomosis somewhere on this otherwise featureless gastric wall is frequently difficult. Once identified, the cannulation with torquable guidewire and catheter is also more problematic than with oesophageal or pyloric strictures. Once the gastroenterostomy has been crossed, exchange for the balloon catheter is performed as previously described, and the

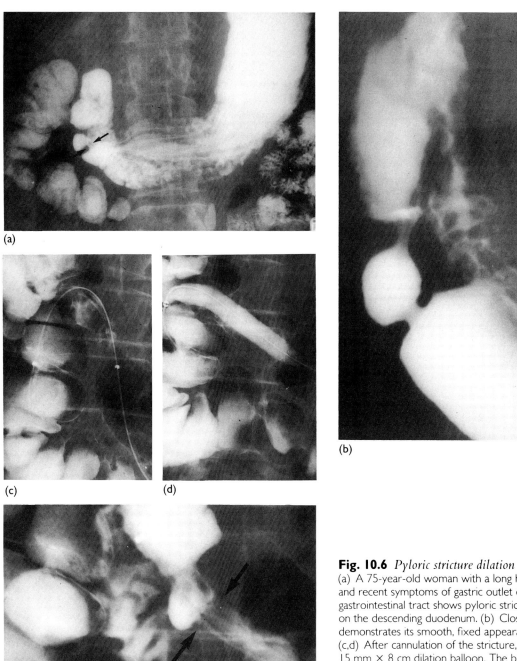

Fig. 10.6 *Pyloric stricture dilation*
(a) A 75-year-old woman with a long history of peptic ulcer disease and recent symptoms of gastric outlet obstruction. Film from upper gastrointestinal tract shows pyloric stricture (arrows) superimposed on the descending duodenum. (b) Close-up view of the stricture demonstrates its smooth, fixed appearance without associated mass. (c,d) After cannulation of the stricture, exchange is made for a 15 mm × 8 cm dilation balloon. The balloon was fully expanded. Note straightening of the bowel segments by the inflated balloon. (e) Following balloon deflation, significant improvement is demonstrated at the dilation site (arrows). This produced complete relief of gastric outlet obstruction symptoms.

anastomosis dilated to 15 mm.

Patients who have strictures following surgery for morbid obesity represent a special category. There are many different operations which are performed for the treatment of obesity, some of which are simply restrictive procedures which reduce the volume of the gastric pouch, and some which combine restriction of the gastric pouch with bypassing of some of the absorptive surface of the jejunum. Gastric pouch restriction is performed most commonly with either vertical or horizontal band gastroplasty. The bypass component, if used, is a gastrojejunal anastomosis. In cases of gastric outlet obstruction following gastroplasty, it is important not to overdilate the strictured surgically created channel, as this could defeat the purpose of the original operation (Menguy, 1990). It is important to consult with the operating surgeon as to the size of the channel that was created. In general, the channel remaining after restrictive gastroplasty should not be dilated to more than 12 mm. Similarly, the anastomosis leading from a restricted gastric pouch through to the jejunal segment should not be overdilated as this would be more likely to allow the patient to overeat by having the gastric pouch empty too readily, thereby defeating the purpose of the original operation (Mishkin *et al.*, 1988) (Fig. 10.7).

There are some instances in which the stomach or gastric remnant will become so dilated and the pylorus or gastroenterostomy so stenotic that the previously described fluoroscopically guided technique will be unsuccessful. Failures can result from an inability to identify the anastomosis, or an inability to advance a catheter over the guidewire once the anastomosis has been traversed. In these cases, strictures can often be successfully identified and cannulated by combining the advantages of fluoroscopically guided and endoscopically guided techniques. An endoscope is advanced to the proximal end of the pyloric or anastomotic stricture and the guidewire advanced into the stricture. However, it is important to monitor the procedure fluoroscopically as well as it is necessary to have a sufficient length of guidewire distal to the stricture, a feature which cannot be appreciated endoscopically. It is not unusual for endoscopists to pass too short a length of stiff guidewire beyond a strictured segment, resulting in buckling back into the stomach when an attempt is made to advance a catheter over the guidewire. Once sufficient wire has been passed, the procedure can continue as previously described. In rare cases, the endoscope may be needed to support the balloon catheter physically as it is pushed through the stricture in order to prevent it from buckling back into the stomach (McLean and Meranze, 1989).

Enteroenteric strictures
The challenge in treating enteroenteric strictures is in getting there. In general, when enteroenteric anastomoses develop strictures, the more distal the anatomic location of the stricture, the less likely it is that the fluoroscopically directed approach will succeed. Torque control is lost with increasing lengths of the path through which the torque is applied and with each turn in the path. After several hundred centimetres and multiple turns, even the most torquable guidewires and catheters simply do not respond in a useful fashion. This is another setting in which the combined endoscopic and fluoroscopic approach may be useful as the endoscopist can sometimes provide access to lesions that would otherwise be untreatable by using a long

endoscope which is guided both with direct visualization and fluoroscopy. Once a wire has been passed through the stricture, the procedure continues as previously described. The balloon size is chosen to approximate but not to exceed that of normal adjacent bowel. The optimal size is 15 to 20 mm (McLean, 1990).

Colorectal strictures
The techniques used for crossing strictures of the lower gastrointestinal tract are essentially the same as those described previously for the upper gastrointestinal tract, including the use of large bore torquable fluoroscopically guided tubes or endoscopically assisted procedures for reaching and cannulating strictures at some distance from the anus or stoma used for access (Figure 10.8).

The main difference in treating strictures of the colon and rectum is the minimum calibre required to produce a normal or near normal faecal stream. The colon is much larger in calibre than the upper gastrointestinal tract and unlike anastomoses in the upper gastrointestinal tract where restriction of flow is intended by the original surgical procedure, this is not the case for colonic anastomoses. The generally available balloon catheters achieve a maximum 20 mm diameter. This sometimes is not sufficient to restore normal defaecation. Therefore, in some cases it is necessary to use more than one balloon simultaneously to achieve an adequate calibre at the strictured site. Most commonly this is accomplished with two balloons side by side, achieving a final diameter that will be less than the sum of both, but greater than that of either (de Lange and Shaffer, 1991). The two-balloon technique produces an oval-shaped dilated segment. Some authors have encouraged the use of three balloons to produce a more circular lumen (Gaylord *et al.*, 1988). The guidelines for the extent of dilation are as outlined previously with the short-term limitation being patient discomfort and the long-term goal being to achieve a lumen as close to that of the adjacent normal bowel as possible, or restoration of normal bowel habits. The size of the balloons chosen will obviously vary significantly, depending on patient age, size of adjacent normal bowel and details of the surgical anastomosis being dilated (Fig. 10.9).

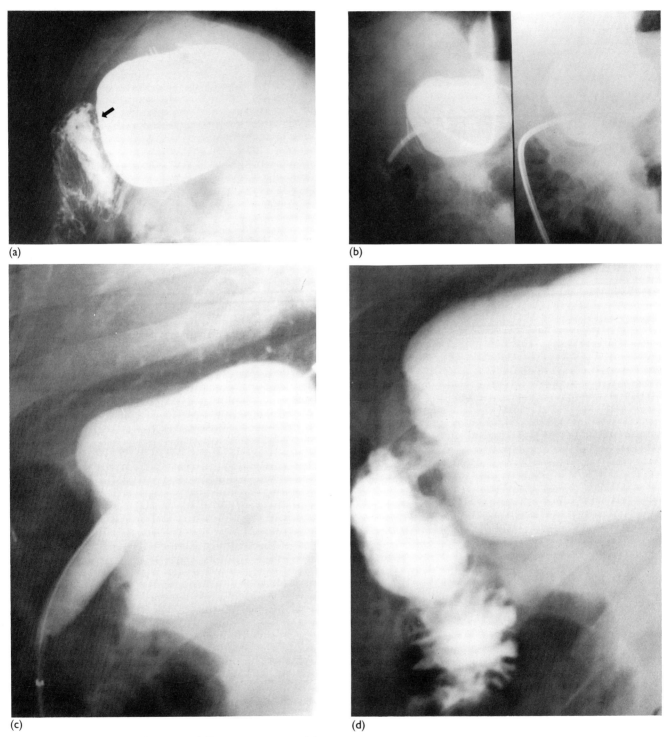

(a)

(b)

(c)

(d)

Fig. 10.7 *Gastric outlet obstruction following gastrojejunal bypass*
A 37-year-old man developed symptoms of gastric outlet obstruction 4 months after gastrojejunal bypass. Two previous endoscopic attempts at cannulation were unsuccessful. (a) Following administration of oral contrast medium, the pinhole of remaining lumen is seen in the centre of the ring of metal anastomotic staples (arrow). Note the steep right posterior oblique projection which best demonstrates the anastomosis. (b) After a brief unsuccessful attempt with the catheter and torquable guidewire, exchange was made for a Herlinger tube which was then successfully manoeuvred through the stricture and into the jejunum. Note how the large calibre of this tube and its torque cable allow transmission of directional force to the tip, even though working through an angle greater than 90°. (c) The Herlinger tube was removed after passing a long exchange wire and a 12 mm balloon catheter advanced and fully expanded. Note how the diameter of the balloon matches that of the ring of surgical staples (arrows). (d) A follow up postdilation contrast medium study demonstrates that the lumen has been restored to the limited diameter intended by the weight reduction surgery. (a)–(c) were reproduced from McLean and Meranze (1989) with permission from *Radiology*

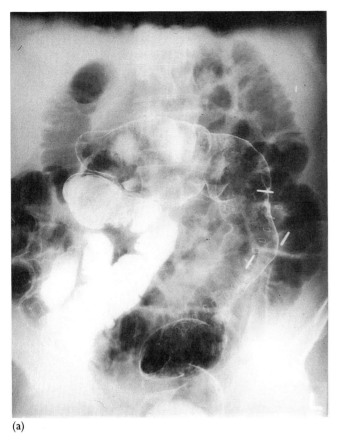

(a)

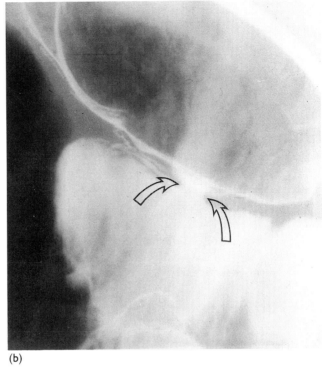

(b)

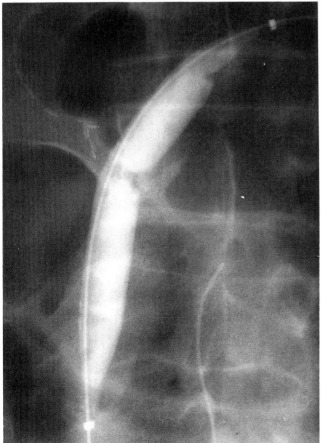

(c)

Fig. 10.8 *Strictured ileocolic anastomosis*
A 65-year-old man had right hemicolectomy for treatment of
adenocarcinoma of the colon. He developed abdominal pain and a
bloating sensation. (a) AP view and (b) close-up view of barium
enema demonstrating strictured ileocolic anastomosis (arrows).
(c) After a colonoscope was used to pass a guidewire through the
stricture, the colonoscope was removed and exchange made for an
initial 15 mm × 8 cm balloon which was expanded to full diameter
(d) without pain. (e) Subsequent exchange for a 20 mm balloon was
made which shows a waist on initial inflation and then (f) expands to
full diameter which produced mild discomfort. (g) Following the
20 mm dilation, a close-up view demonstrates a significant increase
in anastomotic lumen (arrowheads). The patient was asymptomatic
for the subsequent 2 years at which point he died of liver
metastases. (a) was reproduced from McLean (1990) with
permission from *Current Problems in Diagnostic Radiology*. (b–d)
were reproduced from McLean *et al.*, (1987) with permission from
Radiology.

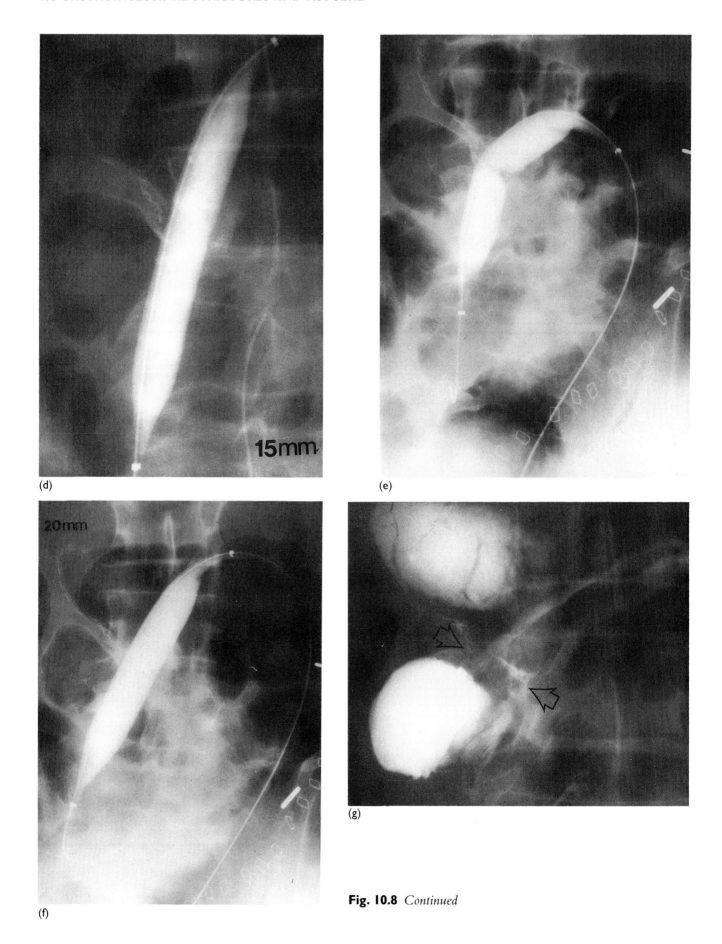

(d)

15mm

(e)

20mm

(f)

(g)

Fig. 10.8 *Continued*

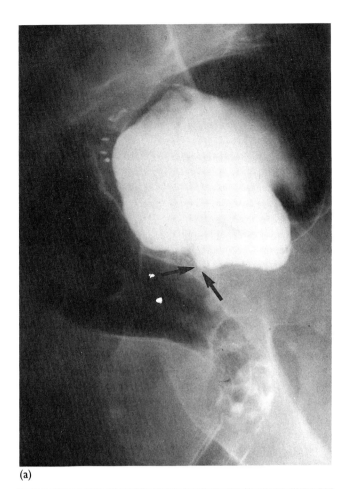

(a)

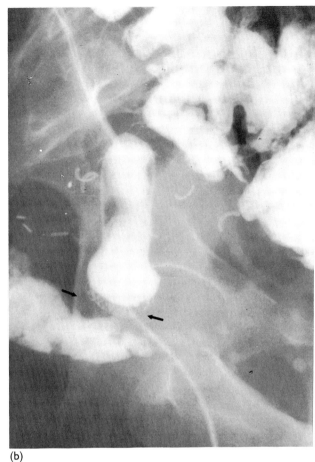

(b)

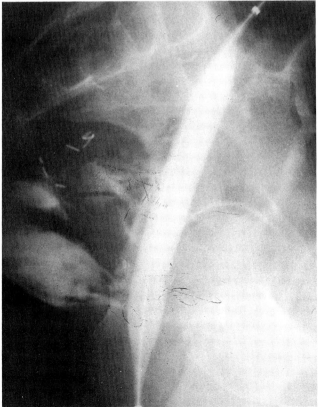

Fig. 10.9 *Balloon dilation of anastomotic rectal stricture*
A 60-year-old woman had sigmoid colon cancer treated with
resection and colorectal anastomosis 1 year before presenting with
nausea and abdominal distention. She had noted the development of
pencil-like stool over the past several months. (a) Lateral view from
barium enema demonstrates a very high-grade stricture at colorectal
anastomosis (arrows). (b) A torquable catheter and soft-tipped
guidewire were used to cannulate the stricture. The location of the
lumen was well marked by the ring of surgical staples (arrows).
(c) The initial 15 mm balloon dilation produced some discomfort, so
no further dilation was performed at that session. (d) Although
some improvement was noted, the thin-calibre stool persisted and
the patient returned for subsequent dilation with a 20 mm balloon.
(e) Repeat contrast enema demonstrates widely patent
anastomosis. The patient's bowel function normalized. Arrows
represent diameter of structure site after dilation. (a) was
reproduced from Meranze (1990) with permission of Gower
Medical Publishing. (b,c) were reproduced from McLean and
Meranze (1989) with permission from *Radiology*.

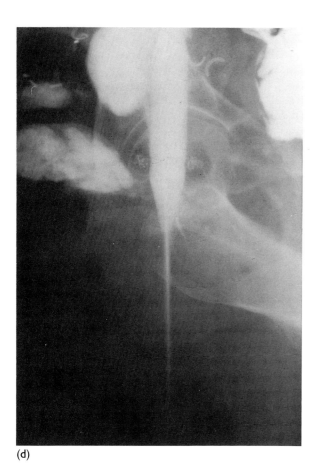

(d)

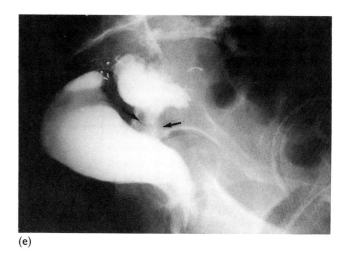

(e)

Fig. 10.9 *Continued*

RESULTS

Upper gastrointestinal strictures

In contrast to oesophageal strictures, relatively small numbers of nonoesophageal gastrointestinal stricture dilations have been published. However, sufficient information and experience have been gained to identify factors which influence success, failure and complications.

TECHNICAL SUCCESS

Gastric and pyloric strictures
Technically, the most difficult part of any gastrointestinal stricture dilation is the intubation of the stricture itself. Once crossed, exchange for the balloon catheter and expansion of the balloon are straightforward in comparison. Strictures of the intact stomach, including those of the body antrum and pylorus, present somewhat greater intubating problems than oesophageal strictures because of the increased length and angulation of the catheter path

before reaching the stricture. However, the angles are fairly gentle and the anatomy predictable, so that the experienced interventionist will succeed in crossing strictures in the intact stomach about 90% of the time compared to a near 100% success rate with oesophageal strictures (Fig. 10.10) (McLean *et al.*, 1987a,b). Similar success would be expected when intubating strictures following horizontal and vertical restrictive gastroplasties.

Gastroenteric strictures
Gastroenterostomies are much more problematic to intubate, with several factors contributing to the difficulties: the anastomoses are variable in location, are usually associated with markedly dilated gastric remnants at the time of presentation, and are often constructed in nondependent locations on the gastric wall. It is not unusual to be unable to identify the actual small bowel lumen with oral contrast medium administration. Fortunately, that in itself does not preclude a successful outcome, but often requires

Fig. 10.10 *Stricture of the body of the stomach following vagotomy and pyloroplasty*
(a) Long stricture in the body of the stomach probably of ischaemic aetiology. (b) A tight waist in a 15 mm balloon at the initiation of dilation. This balloon was fully expanded before (c) exchange for a 20 mm balloon which was also expanded to full diameter without pain. (d) Postdilation contrast medium study demonstrates significant improvement in lumen through the previously strictured site. The patient's symptoms of gastric outlet obstruction resolved. (a,b and d) were reproduced from Meranze (1990) with permission of Gower Medical Publishing.

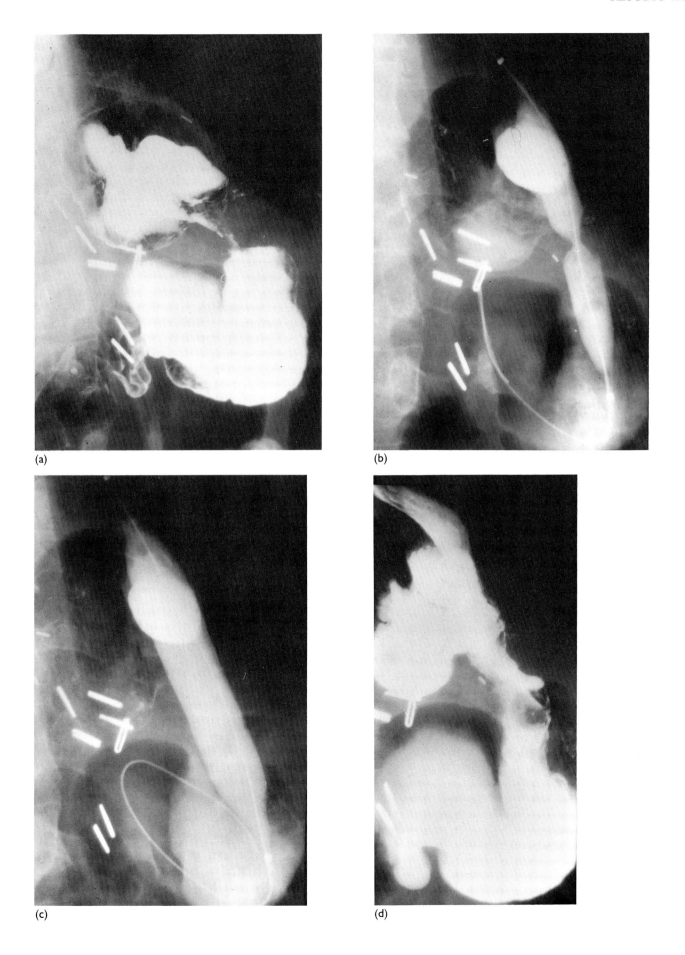

(a)

(b)

(c)

(d)

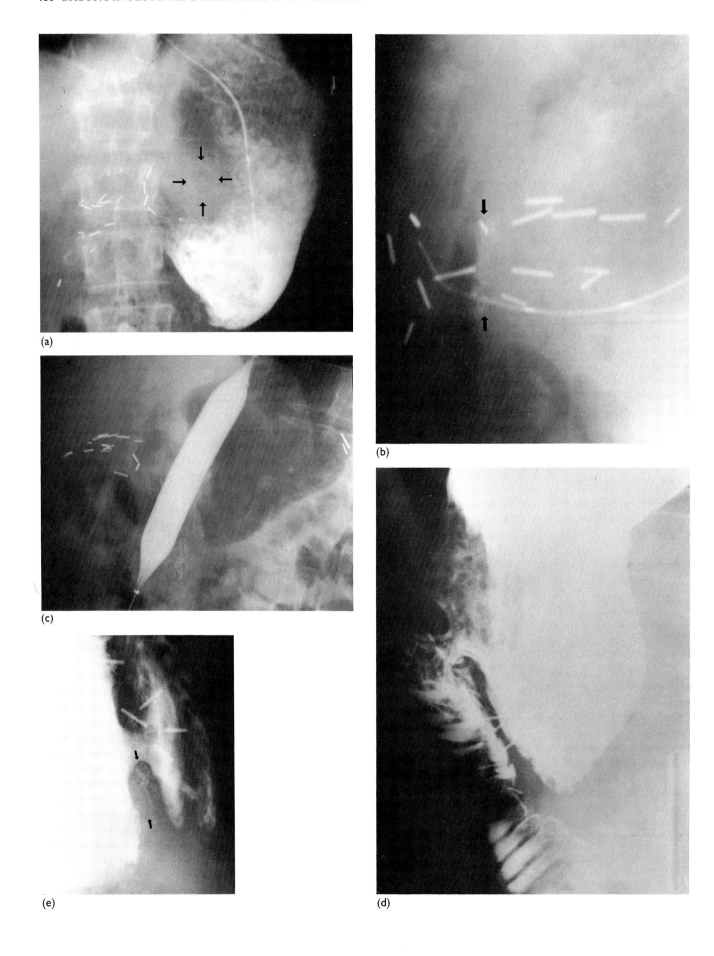

(a)

(b)

(c)

(d)

(e)

extensive searching with torquable guidewires and catheters (Fig. 10.11) (McLean *et al.*, 1987a,b). In addition, once the strictured lumen is located and engaged with the guidewire, the dilated proximal portion of the stomach provides little side support for wires and catheters during advancement across the stenoses. There is another factor which contributes to the reported low success rate in treating gastroenterostomies – the fact that many of the patients are referred after failure of the endoscopic approach. This subselects a group of patients who are very difficult to cannulate by any means, with those patients who are easier to treat never being referred to the interventional radiologist.

The problems encountered with intubation of strictured gastroenterostomies result in a failure rate of intubation of approximately 30%. In a large series, which evaluated a variety of gastrointestinal strictures, no association was found between ease of intubation and stricture characteristics such as diameter, length, eccentricity, irregularity, presence of ulcers or the aetiology of the stricture. The only factors which did correlate with difficulty in intubation were the existence of a surgical anastomosis or failed prior attempts at cannulation (McLean *et al.*, 1987a,b).

CLINICAL SUCCESS

The success of gastrointestinal balloon dilation is determined solely on the basis of clinical improvement. The postdilation appearance, as demonstrated by contrast medium injection, is often very unimpressive (de Lange and Shaffer, 1988). There are several reasons for this: the bowel is a low-pressure, collapsible tube and it is often difficult to produce a

Fig. 10.11 *Successful dilation in the absence of contrast medium demonstration of strictured lumen*
This 55-year-old man had a Billroth II type gastroenterostomy for treatment of severe peptic ulcer disease. (a) Contrast medium instillation through a nasogastric tube demonstrates a dilated stomach containing considerable debris. Although the site of anastomosis is indicated by the ring of surgical staples (arrows) high on the gastric wall, no contrast medium flowed through. (b) After exchange for a Berenstein catheter and Lunderquist–Ring torque guidewire, extensive searching at the site of the anastomotic ring (arrows) resulted in successful cannulation. (c) After advancing the catheter and guidewire well into the jejunum, exchange was made for a 15 mm × 8 cm balloon catheter which was expanded to full diameter. (d) Upper gastrointestinal examination following dilation demonstrates spontaneous flow through the anastomosis.
(e) Close-up view demonstrates the lumen through the ring of surgical staples (arrows). (a–d) were reproduced from Meranze (1990) with permission of Gower Medical Publishing.

maximum diameter with a simple injection of contrast medium. Also, oedema and spasm occur in response to the trauma of balloon dilation, and these contribute to an inconsistent postdilation appearance. The main focus of the contrast medium injection is primarily to look for complications such as extravasation. The end point of dilation, as discussed previously, is determined more by success in achieving the predetermined balloon diameter or reaching a point of patient discomfort.

Pyloric strictures
Patients undergoing dilation for peptic disease in general have been those who are poor surgical candidates. This bias has selected a group of patients whose long-term survival is limited and therefore the long-term evaluation of success of balloon dilation of pyloric strictures is limited. In small numbers of reported cases, success rates of 67–80% have been achieved (Hogan *et al.*, 1986; Kozarek, 1986).

Gastroenterostomies
Fluoroscopically guided balloon dilation of gastroenterostomies produces symptomatic relief in approximately 70% of patients over a 2-year period. As with technical success, stricture irregularity, ulceration and length do not correlate with longterm success or failure. Very eccentric lesions tend to be associated with earlier recurrence of symptoms (McLean *et al.*, 1987b). The problem with eccentricity may be that much of the dilating effect of the balloon is spent on normally distensible bowel wall rather than on the stricture site. Another type of 'lesion' which would be expected to respond poorly is one which is actually a kink in the bowel rather than a true stricture, as sometimes results from poorly positioned loops of bowel involved in gastroenterostomies.

Lower gastrointestinal strictures

TECHNICAL SUCCESS

Compared with the problems encountered in dilating gastroenteric anastomoses, the technical approach to lower gastrointestinal strictures is simpler. Most of the strictures are quite distal so that the course of the catheterization is short, which improves the torque control when steering across the anastomoses. Technical success rates nearing 100% have been consistently reported (McLean *et al.*, 1987a,b; Wilder and Melhem, 1989; de Lange and Shaffer, 1991).

For more proximal colonic lesions or small bowel

to colon anastomoses, the path the catheter has to take can be long and tortuous which limits torquability. It is in these cases that combining the advantages of fluoroscopy and endoscopy may lead to success where either technique alone has failed.

CLINICAL SUCCESS

The goal of balloon dilatation in colonic and rectal strictures is to achieve a normal or near-normal faecal stream. This can be accomplished with a single dilation procedure in about half the patients. Follow-up is relatively short due to the recent development of the procedures, but 2-year symptom-free intervals have been documented. In those requiring multiple dilatations, the symptom-free period between dilatations has been shown to increase progressively. Because access is usually technically easy, multiple repeat dilatations represent a reasonable means of providing long-term symptomatic improvement (de Lange and Shaffer, 1991).

COMPLICATIONS

Few complications from balloon dilatation of gastrointestinal strictures have been reported, and most of these have occurred when treating oesophageal or oesophageal anastomotic lesions (La-Berge *et al.*, 1985; de Lange and Shaffer, 1991). Because the pressures generated within even 'low-pressure' balloons exceed the tensile strength of the bowel wall, it is important that the balloon diameter does not exceed the diameter of the adjacent bowel. Also, care should be taken when the balloon is inflated in a segment of bowel which is not freely mobile: as the balloon is inflated, it also straightens and this can result in an impingement of the end of the balloon catheter against a point in the bowel wall. If that portion of the bowel wall is unable to move away from the balloon, it may perforate. Therefore, it is very important when inflating the balloon to be responsive to the patient's reports of pain and to adjust balloon position accordingly. It has been suggested that balloon dilatation of strictures which occur at previously leaking surgical anastomoses may be more likely to develop leaks after balloon dilatation (de Lange and Shaffer, 1991).

Other than disruption of the bowel wall, the major reported problems with balloon dilatation are bacteraemia and sepsis. During the procedure, there is disruption of the mucosa and bleeding, which may lead to bacteraemia. This is far more likely to occur during manipulations in the lower gastrointestinal tract, where bacteria are normally encountered. However, in patients with gastric stasis, bacterial overgrowth can occur, and it is prudent to treat all patients undergoing balloon dilatation with antibiotics before the procedure.

PERCUTANEOUS MANAGEMENT OF GASTROINTESTINAL FISTULAE

A fistula is defined as an abnormal communication between two epithelial surfaces and can occur from viscus to viscus (internal) or more commonly, viscus to skin (external). Fistulae vary in complexity from a single track to multiple tracks with associated abscesses. In general, the simpler types are more likely to heal with conservative therapy while those associated with abscesses require intervention. Fistulae are divided into low output and high output. Those with outputs of 200 ml or less per 24 hours are considered low and those with greater than 200 ml per 24 hours are high output. The volume of output is inversely related to the likelihood of spontaneous closure without intervention. The location of the fistula along the gastrointestinal tract, and the underlying aetiology, are also important prognostic factors (Ellis and Irving, 1990). Abdominal surgery is the most common cause of gastrointestinal fistulae. Inadvertent enterostomy and leaking anastomoses are the major subgroups of the surgically related fistulae. Penetrating trauma, inflammatory diseases such as ulcers, diverticulitis, radiation enteritis and Crohn's disease as well as malignancy are other common causes.

Gastrointestinal fistulae produce considerable fluid and electrolyte losses. The bowel contents can be quite irritating and quickly lead to skin ulceration. Sepsis and malnutrition frequently complicate the management of fistulae.

DIAGNOSTIC STUDIES

All patients presenting with gastrointestinal fistulae should be evaluated with cross-sectional imaging, preferably CT scanning and contrast medium fistulography.

Computed tomographic scanning

The primary purpose of cross-sectional imaging is to detect abdominal abscesses associated with the fistulae which will need to be drained before healing can occur. Also, CT scanning can produce additional diagnostic information such as demonstrating the characteristic changes of Crohn's disease in the mesentery or evidence of a phlegmonous process related to diverticulitis or pancreatitis. Occasionally, underlying malignancies can be suggested by CT scanning. If an abnormal soft tissue density suggestive of abscess is detected in association with the fistula, it is useful to perform a contrast medium injection of the fistula (see below) followed by re-imaging at the level of the suspected abscess to see if contrast medium has entered it. If contrast medium has entered the abscess, indicating communications with the fistula, it can be drained under fluoroscopic guidance in most cases. If it does not, a separate puncture of the abscess will be required and this should be done under CT guidance while the patient is still on the table. After drainage of the abscess, fluoroscopic evaluation of the fistula is performed.

Contrast medium fistulography

Access to CT scanning is not always available at short notice, and coordination of scanning with fistula track injection is not always practical. Rather than delay the evaluation of the patient, fistulography should be promptly performed when there is an anticipated delay in CT scanning. The purpose of the fistulogram is to determine the site of origin of the fistula to establish whether there is obstruction distal to that site and to show any abscess communicating with the track. This study should be performed initially with the least possible invasion of the track. This is best accomplished with a funnel-shaped 'Christmas tree' adapter attached to a syringe of iodinated contrast medium. The adapter protrudes minimally into the fistula while providing obturation of the entrance which encourages the contrast medium to traverse the fistula rather than

refluxing out to the skin. For studying fistulae with larger openings, Foley catheters are inserted a short distance into the fistula and then the expanded balloon is used to occlude the opening (Fig. 10.12). Blind advancement of catheters too far into fistulae can result in false passages which can make subse-

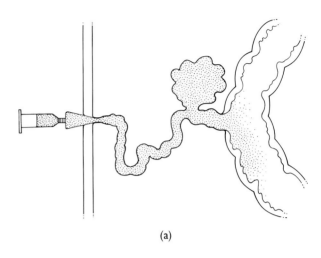

(a)

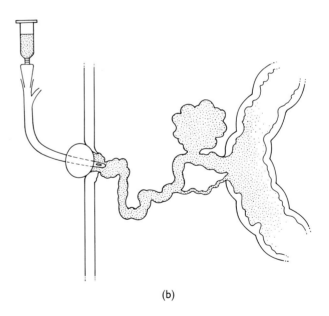

(b)

Fig. 10.12 *Fistulogram technique*
(a) A 'Christmas tree' adapter is applied gently at the fistula opening, and contrast medium is injected. The minimal depth to which the adapter penetrates the track minimizes the risk of dissection. The fistulogram demonstrates the course and calibre of the fistula track as well as the presence of associated abscesses and the patency of the bowel. (b) When skin breakdown widens the skin opening to a fistula, a Foley catheter with a 30 ml balloon is used. The balloon is expanded and then the tip of the catheter is advanced slightly into the track while the balloon catheter obturates the opening.

quent cannulation of the tract difficult, if not impossible. Filming is performed in multiple projections during contrast medium injection to record the course and calibre, to detect the relationship with abscesses and to assess the condition and distal patency of the bowel. If the fistulogram demonstrates no distal obstruction and no associated abscess, conservative treatment has a reasonable chance of success.

CONSERVATIVE MANAGEMENT

Provided the volume of fluid and electrolyte output is not excessive, it is reasonable to attempt a course of conservative management designed to produce spontaneous closure. Conservative treatment focuses on decreasing fistula output while treating the complications of infection, malnutrition and skin breakdown that accompany gastrointestinal fistulae.

Nutrition

Central venous parenteral nutrition is usually required to meet the caloric demands of the healing fistula while keeping the volume of intestinal contents to a minimum. The contents of the parenteral solution can be modified to correct fluid and electrolyte imbalances as well as to provide sufficient calories. In some patients with low output fistulae from the distal small bowel or colon, external nutrition using a low residue elemental diet can provide sufficient calories without significantly increasing fistula output. Adjunctive pharmacologic therapy with somatostatin and H^2 blockers can also contribute further to the reduction of enteric and pancreatic secretion.

Skin care

Gastrointestinal contents can be extremely irritating to the skin, leading rapidly to painful excoriation and even digestion of the abdominal wall. It is imperative to minimize the contact of the fistula contents with the skin. This is most commonly done using a sheet of stoma adhesive (Hollister Inc. Illinois) which is cut to size around the fistula opening and fitted with a flanged ring to which a bag can be attached for collecting the fluid. The standard wafer can be

tailored to the specific patient's anatomy with the addition of Karaya paste. Early involvement of the hospital stoma therapy department for design and management of these skin barriers is extremely important.

Antibiotic therapy

Although most patients who have sepsis associated with fistulae will receive antibiotics for associated abscesses, signs of systemic infection can frequently accompany simple fistulae and should be treated with appropriate intravenous antibiotics. In particular, the antibiotics may be useful for promoting healing of damaged, inflamed skin at the fistula site.

The overall success of conservative treatment should be reassessed after 2 weeks. If significant improvement has occurred, it is reasonable to continue conservative therapy until a point of no progress is encountered or spontaneous closure occurs. Conservative therapy will not succeed if there is loss of continuity of the bowel or distal obstruction, or sometimes if there is infiltration of the fistula tract by an underlying infectious, inflammatory or malignant process. In any case, if no significant improvement is noted after 2 weeks of conservative measures, more active intervention should be pursued. In most cases, this should initially involve interventional radiological methods with surgery reserved for those conditions which prove resistant to the less invasive methods described in this section.

PERCUTANEOUS INTERVENTION

Enterocutaneous fistulae that are complicated by associated abscesses at the time of presentation, fistulae which are discovered after initial drainage of abscesses, and simple fistulae which do not respond to the previously outlined conservative measures are all amenable to percutaneous therapy. The goals of percutaneous therapy are: (1) to place a tube or tubes precisely which will divert the bowel contents away from the site of the defect in the bowel wall and (2) to drain thoroughly associated abscess collections. The specific techniques used vary depending on whether the enterocutaneous fistula is spontaneous or one discovered after the initial drainage of an abscess, but the principles are the same.

Cannulation of enterocutaneous fistulae

Intubation of the fistula track is not attempted until a thorough fistulogram including filming in multiple projections has been performed. The purpose of this is to delineate the interconnections of the sinus tracks completely and to define as early as possible the relationship between the abscesses, the sinus track and the hole in the bowel. Cannulation of the sinus track must be performed very carefully to avoid dissection of these often friable, tortuous pathways. The initial attempts should be made with a small calibre (8 or 10 Fr) red rubber catheter in which an enlarged side hole has been added. The tip of the tube is coated with a water-soluble lubricant and then gently advanced into the tract. Traction on the skin can frequently reduce sharp angulations in the tract to facilitate catheterization. The blunt, lubricated tip of the catheter will tend to remain within the lumen rather than digging into the walls of the fistula (McLean *et al.*, 1982). If sharp angulations or tight stenoses prevent additional progress with the blunt tipped red rubber catheter, the back end of a guidewire can be angled approximately 45° about 1 cm from the tip and advanced to the end of the red rubber catheter. This will result in angulation of the tip of the catheter as well as stiffening the catheter. By spinning the catheter, some degree of torque can be transmitted to the tip and thereby provide the orientation for negotiating the more abrupt angles. The presence of the wire will also allow the application of greater force without producing an 'accordion' effect on the catheter itself. Once the catheter has reached the lumen of the bowel, additional contrast medium is injected to evaluate the distal patency of the bowel. A guidewire is then advanced through the tube and out the side hole and the red rubber catheter removed.

If cannulation requires more manoeuverability than can be provided by the red rubber catheter, a torquable guidewire and flexible straight or slightly angled catheter are used to steer actively through the tortuous tract. As each bend is traversed by the wire, the catheter is advanced through it and the sequence is repeated until the source of the leak is reached. Extreme care must be taken to avoid making false passages through the fragile wall of the fistula with the tip of the wire. If the wire does not move with a gentle push, a new angle must be chosen. Once the tip is aligned with the channel, it will advance easily with gentle pushing.

Diversion of enteric contents

Control of the effluent from the bowel is achieved in one of two ways. If the hole in the bowel wall is quite large, a Malecot-tipped catheter or T-tube is placed within the bowel lumen to collect and divert the bowel contents before they leak to the adjacent tissues. This method also has some advantage when the leaking fluid is particularly irritating, as seen with gastric or duodenal fistulae which will contain large amounts of hydrochloric acid or pancreatic enzymes. As an alternative, in smaller fistulae, a sump-type drainage tube is positioned immediately outside the hole in the bowel to collect the contents actively as they leak and to divert them away from the adjacent tissues (McLean, 1990) (Fig. 10.13).

In some cases, although obvious enteric contents are leaking out through the skin, injection of the sinus track does not delineate bowel mucosa, but fills one or more irregular spaces which are the associated abscesses. Each of these recesses or abscesses should be gently explored with a torquable catheter and soft-tipped guidewire, such as a Bentson guidewire until the communication with the bowel is found. It is important to be careful not to probe the tracks or abscesses too vigorously as it is easy to dissect into the inflamed and friable tissues which can both complicate the picture by creating false passages as well as produce massive bacteraemia which can induce septic shock. If no communication to the bowel is apparent after thorough but gentle exploration, a sump drainage tube of the largest calibre the track will accommodate (16 to 24 Fr) is placed centrally within the track or in the largest of the abscess collections. After several days of drainage, injecting contrast medium through the tube will delineate the fistula which can then be cannulated and controlled.

Drainage of abscesses

The second major focus of percutaneous therapy is drainage of the associated abscesses. This may require multiple drainage tubes. Although the principles for treating these abscesses are the same as for abscesses not associated with fistulae, in general they require more aggressive and longer term use of sump catheters. It can be difficult in many cases to place multiple, relatively large bore drainage tubes through a spontaneous fistula track or through the access route of an initially drained abscess, even with the use of very stiff exchange guidewires. Sometimes it is necessary to settle for less than the ideal calibre at the initial drainage site with gradual track dilation and increase in size of drainage tubes over the next several days.

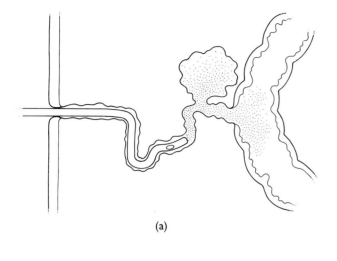

(a)

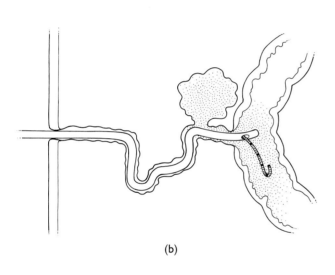

(b)

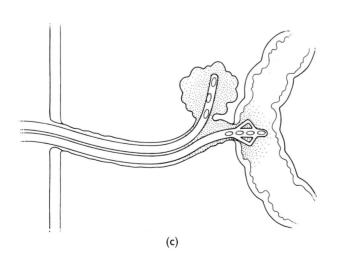

(c)

Drainage tube management

Once the fistula is controlled, the abscesses will heal with tube drainage. As with any drainage tube, these tubes must be kept patent to be successful. This requires frequent tube flushing, manipulation and replacement. Repositioning and replacement are done in conjunction with repeat sinograms which monitor progress and direct optimal catheter positioning. When sump tube output falls below 100 ml a day, the increased frequency of occlusion of the lumen generally outweighs the advantages of active drainage and the sump tubes are then replaced with single lumen gravity drainage tubes. When the sinogram demonstrates resolution of abscess cavities, the abscess tubes are backed out incrementally and removed. Once the abscesses are resolved, there remains a single fistula track drain with a single tube. Depending on the volume of output, this can be a sump tube (greater than 100 ml) or single-lumen tube (less than 100 ml per 24 hours). The tube is positioned several centimetres from the bowel wall. When drainage ceases, a final sinogram is performed to confirm fistula closure and the tube is once again removed incrementally (Fig. 10.14).

Fig. 10.13 *Cannulation of a fistula*
(a) The blunt-tipped, flexible red rubber catheter is the ideal tool for cannulating tortuous fistulae. The blunt tip minimizes the risk of dissection while the body of the catheter conforms readily to the course of the sinus track. (b) After the catheter has reached the bowel or associated abscess, a torquable guidewire is advanced through the catheter and out the enlarged side hole. (c) Once a stiff guidewire traverses the tract, the tract will straighten. Using the coaxial technique, the track is dilated and drainage tubes are put in place. In this example, the Malecot catheter diverts the bowel contents while the sump catheter drains the abscess. In some cases a single catheter will perform both functions and in other cases multiple drainage tubes external to the bowel will be required.

Fig. 10.14 *Entercutaneous fistula following appendectomy*
This 25-year-old man had an appendectomy 2 weeks before presentation. The appendix was gangrenous and had leaked. Patient presented with sepsis and leaking of enteric contents·from the suture line. (a) The initial fistulogram demonstrates an amorphous contrast medium collection with faint visualization of bowel. Because of the sepsis, no further manipulation was performed after initial cannulation. (b) After 3 days of drainage, there is decrease in size of the abscess cavity with several fistula tracks to irregular, oedematous small bowel. A single sump drain could be placed within the abscess and adjacent to the dominant fistula. (c) After 10 days this was exchanged for a red rubber catheter. Injection of this demonstrates near complete resolution of the cavity and a smooth single communication with more normal appearing small bowel. (d) After two additional weeks of outpatient drainage, contrast medium injection demonstrates a simple tube tract with resolution of the fistula. The tube was removed.

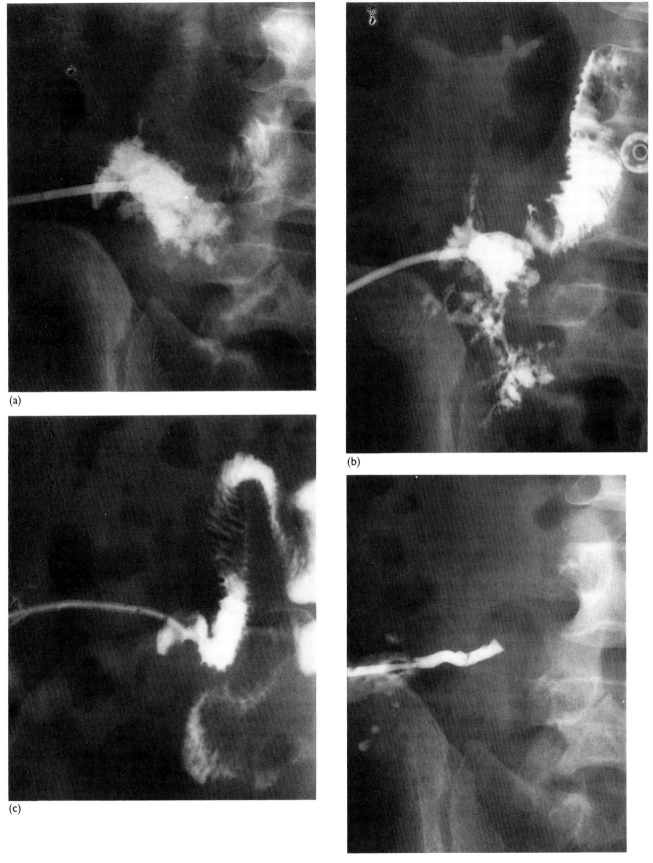

(a)

(b)

(c)

(d)

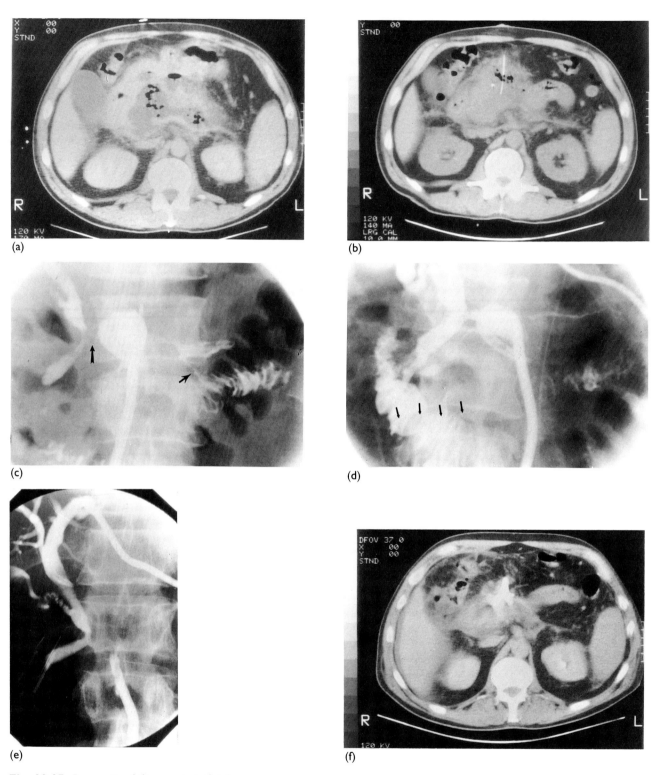

(a)　(b)　(c)　(d)　(e)　(f)

Fig. 10.15 *Pancreatico-biliary-enteric fistula*

This 55-year-old man developed fulminant pancreatitis following diagnostic ERCP. (a) An abdominal computed tomographic (CT) scan performed to evaluate sepsis demonstrates large pancreatic abscess. (b) Computed tomographic-guided puncture of the abscess was performed and then exchange made for a sump drainage tube. (c) Abscess catheter injection 2 weeks after drainage demonstrates fistulae to the common bile duct (arrow) and distal duodenum (arrowhead). (d) With continued drainage, the small bowel fistula healed, but the common bile duct fistula persisted. Note also the pancreatic duct filling (arrows). (e) The biliary contents were diverted using transhepatic biliary drainage after an additional 4 weeks of single tube drainage failed to heal the fistula. Two weeks following transhepatic biliary drainage, the biliary and pancreatic fistulae healed. (f) Final CT scan also demonstrates resolution of the abscess, and the tube was removed incrementally 3 months after initial drainage.

Gastrointestinal fistulae related to pancreatic inflammatory disease are particularly challenging as abnormal communications with the pancreatic duct and biliary tree are common. The fistulae can be spontaneous (pancreatico-enteric-cutaneous), but more commonly they are found after pancreatic surgery or drainage of pancreatic abscesses. Catheter diversion of both pancreatic and biliary output is sometimes necessary to allow healing. Prolonged drainage is routinely required in this setting (Freeny et al., 1988). However, the principles applied are the same as for other gastroenteric fistulae with the initial focus being resolution of the abscesses followed by closure of the fistulae (Fig. 10.15).

RESULTS

Before the aggressive use of total parenteral nutrition, the spontaneous closure rate was reported to be less than 25%. Parenteral nutrition has approximately doubled the spontaneous closure rate, but at least 50% of fistulae will require intervention (MacFadyen et al., 1973; Thomas, 1981).

The success rates reported for complete closure of gastrointestinal fistulae are 60–80% using percutaneous techniques (McLean et al., 1982; Papanicolaou et al., 1984; Kerlan et al., 1985, 1991; Lambiase et al., 1989). The breakdown of patient characteristics by aetiology, location and fistula volume varies among the published reports. Nevertheless, these studies have clarified the factors which influence the success of percutaneous treatment.

Fistula output

In contrast to earlier surgical series, the success rates for curing high output and low output fistulae by percutaneous methods are comparable, with low output fistulae resolving in 63–84% and high output fistulae in 52–100% of published cases. The explanation is a more recent emphasis on parenteral nutrition and a more effective control of effluent by a combination of bed rest, cimetidine, somatostatin and precise drainage tube placement. More recent surgical reports also suggest that initial output volume is a less significant factor than it was historically (Hollender et al., 1983; Geerdsen et al., 1986).

Fistula location

In both the surgical and the radiological literature, gastric and proximal duodenal fistulae are very difficult to close without surgery (Ellis and Irving, 1990; La Berge et al., 1992). The fistulae are usually high in output and result from breakdown of surgical anastomosis. The high volume of irritating effluent causes severe fluid and electrolyte problems as well as excoriation of the skin. Overall, fewer than 25% of gastric fistulae will close without surgery (Figs. 10.16 and 10.17).

Distal duodenal and other small bowel fistulae tend to close at an 'average' rate (60–80%) consistently within the ranges reported in the literature. Most duodenal and small bowel fistulae are postoperative complications and this group includes duodenal stump leaks and other afferent limb leaks following gastroenterostomy.

The success with colonic fistulae in reported series varies widely from about 10 to 70%. Colonic effluent is thick and particulate, requiring large bore tubes and intensive tube maintenance regimens (Fig. 10.18). One possible explanation for the variable success rates may be the high rate of spontaneous closure with conservative measures. Early percutaneous intervention in a patient with a colonic fistula may be credited with success which would have occurred eventually, although over a longer period, without it. On the other hand, if percutaneous drainage is used only after more conservative measures have failed to produce spontaneous closure, the success rate attributed to percutaneous drainage will be considerably lower.

Fistula aetiology

The aetiology of a fistula clearly affects the prognosis for closure. In general, if the underlying bowel at the site of the fistula is diseased, the likelihood of cure is small. Crohn's disease is a good example. In reported series, spontaneous fistulae to bowel involved with Crohn's disease closed with tube drainage in only 16% of cases (Fig. 10.19) (Casola et al., 1987; Safrit et al., 1987; Doemeny et al., 1988; Lambiase et al., 1988). However, in these series, postoperative fistulae in patients with Crohn's disease, i.e. after the removal of the diseased segment, closed 80% of the time, behaving more like standard postoperative fistulae. Fistulae involving segments of bowel affected by radiation enteritis or colitis would not be expected to close spontaneously. When malignancy is the underlying

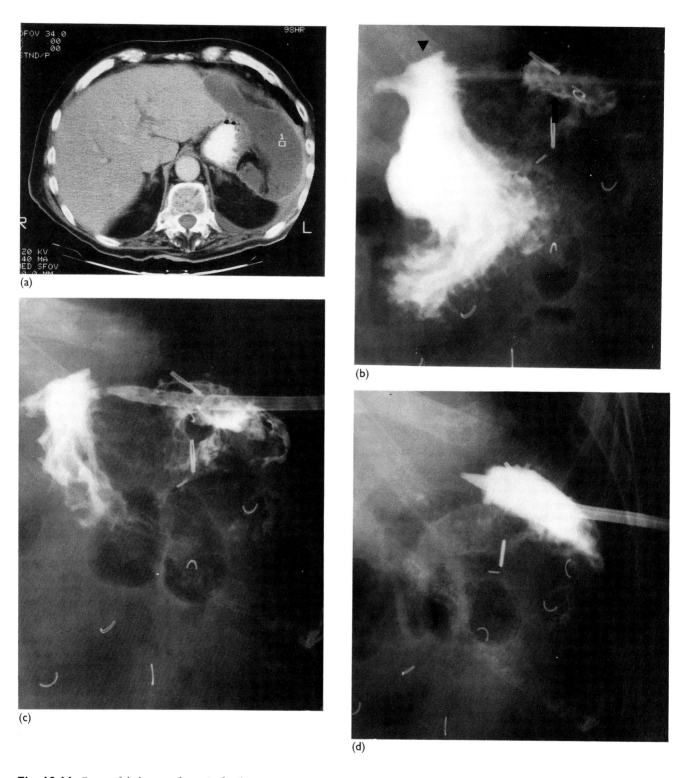

Fig. 10.16 *Successful closure of gastric fistula*
A 65-year-old man underwent left colonic resection which included splenectomy. He presented 1 week after surgery with fever and leukocytosis. Purulent fluid was leaking from the previous surgical drain site. (a) Initial abdominal computed tomographic (CT) scan demonstrates a large left upper quadrant abscess. (b) Injection of the fistula track filled both the abscess (arrow) and stomach (arrowhead). The red rubber catheter advanced easily into the stomach. Inadvertent intraoperative gastric injury no doubt led to leakage of the gastric contents into the splenic bed. (c) A Ring–McLean (Cook Inc., Bloomington, Indiana) sump was placed initially and positioned with side holes at the gastric fistula as well as within the abscess. The spot film after 10 days with some drainage demonstrates a small residual abscess cavity. (d) After two additional weeks of drainage, the tube injection demonstrates closure of the gastric fistula and a smooth walled minimal residual abscess cavity which resolved completely with an additional week of straight catheter drainage.

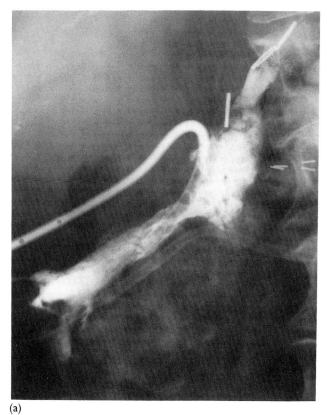

(a)

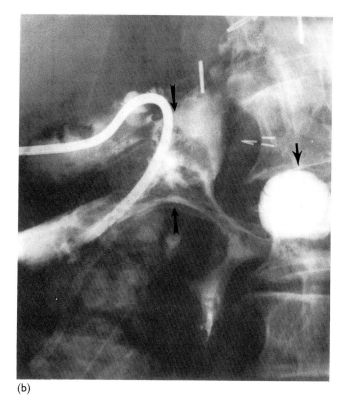

(b)

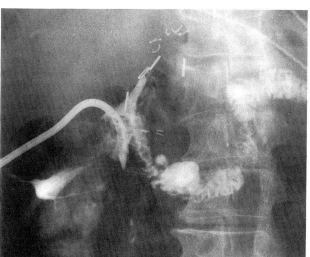

(c)

Fig. 10.17 *Control of proximal duodenal fistula without closure*

A 70-year-old man with severe obstructive pulmonary disease underwent oesophagogastrectomy for cancer of the oesophagus. The surgery included mobilization of the distal stomach and proximal duodenum to enable gastric pull-through and reanastomosis with the oesophagus. The patient presented with sepsis and leaking of enteric contents from a right flank drain site 10 days after surgery. (a) The initial fistulogram demonstrated an irregular collection in Morrison's pouch which was drained initially with a sump catheter. (b) Abscess tube injection 3 days later after the patient had stabilized demonstrates large communication with the duodenum (arrows). A large diverticulum is seen in the third portion of the duodenum (arrowhead). (c) The sump catheter was positioned immediately adjacent to the large duodenal hole and sump drainage was maintained for 3 months. Because the patient required chronic steroid therapy to manage his pulmonary disease, healing was retarded in this high output fistula. Catheter drainage produced resolution of the Morrison's pouch abscess and control of the effluent, but closure of the fistula had not occurred 4 months after surgery when the patient died of pulmonary complications.

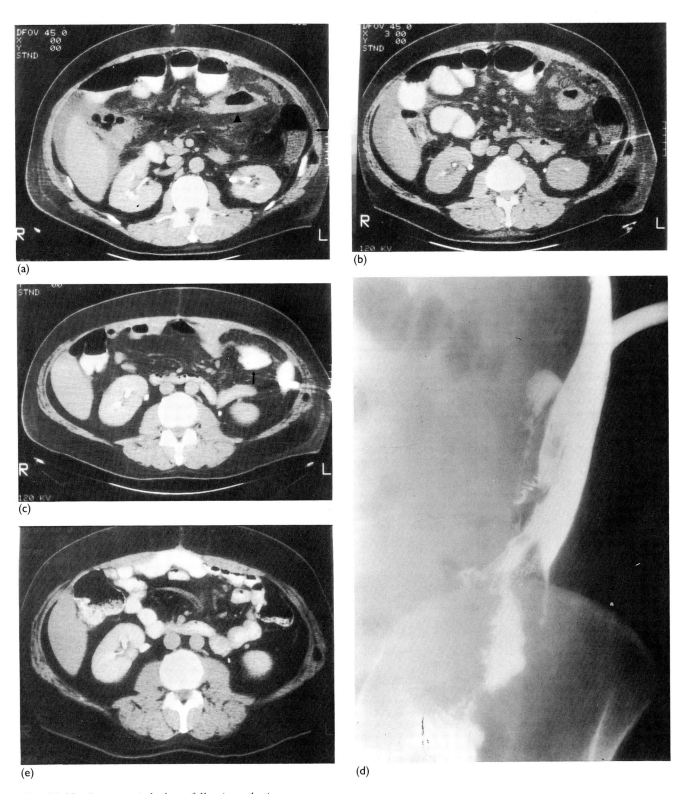

Fig. 10.18 *Anastomotic leakage following colonic surgery*

A 56-year-old man underwent distal transverse colectomy for diverticular bleeding, unresponsive to catheter therapy. One week after surgery he developed sepsis and extraluminal gas was noted on an abdominal film. (a) Initial abdominal computed tomographic (CT) scan demonstrates ascites, abscess with air fluid level in the left flank (arrow) and abnormal thick walled bowel (arrowhead). (b) Because there was no cutaneous fistula, CT-guided puncture of the left flank abscess was performed. (c) Drainage was initiated with a 16 Fr sump catheter which is seen within the smaller abscess cavity on a 1 week follow-up CT scan. The abscess filled spontaneously from oral contrast medium given for this examination. Note clearing of ascites since time of presentation and improvement in appearance of adjacent bowel (arrow). (d) Sump drainage tube size was gradually increased to 24 Fr to handle the thick particulate colonic effluent. Catheter injection 2 weeks after initial drainage demonstrates large anastomotic leak. This fistula closed over the following 2 weeks, but required frequent tube irrigation and exchange to maintain patency. (e) A CT scan to evaluate incisional hernia 2 months after drainage tube removal demonstrates complete clearing of the abscess cavity and normal appearance of adjacent bowel.

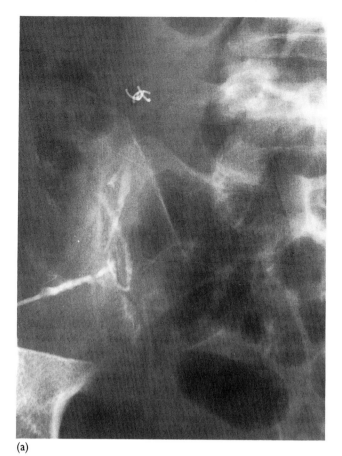

(a)

(b)

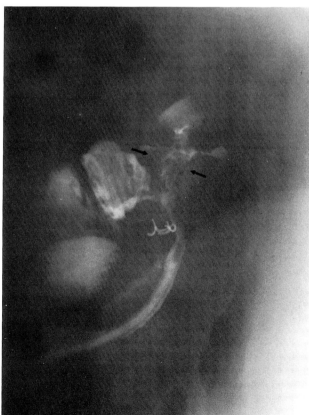

(c)

Fig. 10.19

A 26-year-old woman with Crohn's disease presented with fever and spontaneous drainage from cutaneous fistula. (a) Fistulogram through a 'Christmas tree' adapter demonstrated several sinus tracks, but no bowel filling. (b) After cannulation with a red rubber catheter, contrast medium injection shows fistulae (arrow) associated inflammatory change in the bowel (arrowhead), and small abscess cavity (open arrow). (c) Six weeks of drainage healed the abscess and reduced the multiple tracks to a single fistula to the inflammed segment of bowel (arrows). The abnormal bowel was removed surgically. (a–c) were reproduced from Burke (1990) with permission of Gower Medical Publishing.

cause of fistulization, cure without surgery is rare. In contrast, fistulae secondary to curable inflammatory processes like diverticulitis (Neff *et al.*, 1987) and appendicitis (Jeffrey *et al.*, 1988) have responded well to tube drainage in reported cases.

Fistulae secondary to inflammatory pancreatic disease demonstrate decreased cure rates in some series (Lambiase *et al.*, 1989), but not in others (Freeny *et al.*, 1988). It is possible that the success rate is affected by the duration of tube drainage allowed and calibre of tubes used before the decision to operate. It is clear that these fistulae require the longest duration of drainage with 3 months of tube drainage being common before complete resolution.

REFERENCES

BURKE DR (1990). Percutaneous abscess drainage. In: *Atlas of Interventional Radiology*. Edited by Cope C, Burke DR, Meranze SG. Gower Medical Publishing, New York, London.

CASOLA G, VAN SONNENBERG E, NEFF CC, SABA RM, WITHERS C, EMARINE CW (1987). Abscesses in Crohn's disease: Percutaneous drainage. *Radiology* **163**: 19–22.

DE LANGE EE, SHAFFER HA JR (1988). Anastomotic strictures of the upper gastrointestinal tract: Results of balloon dilatation. *Radiology* **167**: 45–50.

DE LANGE EE, SHAFFER HA JR (1991). Rectal strictures: Treatment with fluoroscopically guided balloon dilation. *Radiology* **178**: 475–9.

DOEMENY JM, BURKE DR, MERANZE SG (1988). Percutaneous drainage of abscesses in patients with Crohn's disease. *Gastrointestinal Radiology* **13**: 237–41.

ELLIS H, IRVING M (1990). Gastrointestinal and biliary fistulae. In: *Maingot's Abdominal Operations* Volume I, pp. 315–34. Edited by Schwartz SI, Ellis H. Appleton & Lange, Norwalk, Connecticut/San Mateo, California.

FREENY PC, LEWIS GP, TRAVERSO LW, RYAN JA (1988). Infected pancreatic fluid collections: Percutaneous catheter drainage. *Radiology* **167**: 435–41.

GAYLORD GM, PRITCHARD WF, CHUANG VP, CASARELLA WJ, SPRAWLS P (1988). The geometry of triple balloon dilatation. *Radiology* **166**: 541–5.

GEERDSEN JP, PEDERSEN VM, KJAERGARD HK (1986). Small bowel fistulas treated with somatostatin: Preliminary results. *Surgery* **100**: 811–14.

HERRINGTON J, SAWYERS J (1990). Complications following gastric operations. In: *Maingot's Abdominal Operations*, Volume I, pp. 701–30. Edited by Schwartz SI, Ellis H. Appleton & Lange, Norwalk, Connecticut/San Mateo, California.

HOGAN RB, HAMILTON JK, POLTER DE (1986). Preliminary experience with hydrostatic balloon dilation of gastric outlet obstruction. *Gastrointestinal Endoscopy* **32**: 71–74.

HOLLENDER LF, MEYER C, AVET D, ZEYER B (1983). Postoperative fistulas of the small intestine: Therapeutic principles. *World Journal of Surgery* **7**: 474–80.

JEFFREY RB, FEDERLE MP, TOLENTINO CS (1988). Periappendiceal inflammatory masses: CT-directed management and clinical outcome in 70 patients. *Radiology* **167**: 13–60.

KERLAN RK JR, GORDON RL, RING EJ (1992). Non-operative management of enteric fistulas: Results in 53 patients. *Journal of Vascular and Interventional Radiology* (in press).

KERLAN RK, JEFFREY RB, POGANY AC, RING EJ (1985). Abdominal abscess with low-output fistula: Successful percutaneous drainage. *Radiology* **155**: 73–5.

KOZAREK RA (1986). Hydrostatic balloon dilatation of gastrointestinal stenoses: A national survey. *Gastrointestinal Endoscopy* **32**: 15–19.

LA BERGE JM, KERLAN RK, JR, GORDON RL, RING EJ (1992). Nonoperative Treatment of Enteric Fistulas: Results in 53 patients. *Hepatic, Biliary and Gastrointestinal Intervention* **3**: 353–7.

LA BERGE JM, KERLAN RK JR, POGANY AC, RING EJ (1985). Esophageal rupture: Complication of balloon dilatation. *Radiology* **157**: 56.

LAMBIASE RE, CRONAN JJ, DORFMAN GS, PAOLELLA LP, HAAS RA (1988). Percutaneous drainage of abscesses in patients with Crohn's disease. *American Journal of Roentgenology* **150**: 1043–5.

LAMBIASE RE, CRONAN JJ, DORFMAN GS, PAOLELLA LP, HAAS RA (1989). Postoperative abscesses with enteric communication: Percutaneous treatment. *Radiology* **171**: 497–500.

LONDON RL, TROTMAN BW, DIMARINO AJ JR, OLEAGA JA, FREIMAN DB, RING EJ, ROSATO EF (1981). Dilatation of severe esophageal strictures by an inflatable balloon catheter. *Gastroenterology* **80**: 173–5.

MACFADYEN BV, DUDRICK SJ, RUBERT RL (1973). Management of gastrointestinal fistulas with parenteral hyperalimentation. *Surgery* **74**: 100–5.

MCLEAN GK (1990). Interventional radiology of the gastrointestinal tract. *Current Problems in Diagnostic Radiology* 107–8.

MCLEAN GK, COOPER GS, HARTZ WH, BURKE DR, MERANZE SG (1987a). Radiologically guided balloon dilation of gastrointestinal strictures. *Radiology* **165**: 35–40.

MCLEAN GK, COOPER GS, HARTZ WH, BURKE DR, MERANZE SG (1987b). Radiologically guided balloon dilation of gastrointestinal strictures. *Radiology* **165**: 41–3.

MCLEAN GK, LEVEEN R (1989). Shear stress in the performance of esophageal dilatation: A comparison of balloon dilatation and bougienage. *Radiology* **172**: 983–6.

MCLEAN GK, MACKIE JA, FREIMAN DB, RING EJ (1982). Enterocutaneous fistulae: Interventional radiologic management. *American Journal of Roentgenology* **138**: 615–19.

MCLEAN GK, MERANZE SG (1989). Interventional radiologic management of enteric strictures. *Radiology* **170**: 1049–53.

MENGUY R (1990). Morbid obesity. In: *Maingot's Abdominal Operations*, Volume I, pp. 771–90. Edited by Schwartz SI, Ellis H. Appleton & Lange, Norwalk, Connecticut/San Mateo, California.

MERANZ SG (1990). Gastrointestinal interventions. In: *Atlas of Interventional Radiology*. Edited by Cope C, Burke DR, Meranz SG. Gower Medical Publishing, New York, London.

MISHKIN JD, MERANZE SG, BURKE DR, STEIN EJ, MCLEAN GK (1988). Interventional radiologic treatment of complications following gastric bypass surgery for morbid obesity. *Gastrointestinal Radiology* **13**: 9–14.

NEFF CC, VAN SONNENBERG E, CASOLA G, WITTICH GR, HOYT DB, HALASZ NA, MARTINI DJ (1987). Diverticular abscesses: Percutaneous drainage. *Radiology* **163**: 15–18.

PAPANICOLAOU N, MUELLER PR, FERRUCCI JT, DAWSON SL, JOHNSON RD, SIMEONE JF, BUTCH RJ, WITTENBERG J (1984). Abscess-fistula association: Radiologic recognition and percutaneous management. *American Journal of Roentgenology* **143**: 811–15.

SAFRIT HD, MAURO MA, JAQUES PF (1987). Percutaneous abscess drainage in Crohn's disease. *American Journal of Roentgenology* **148**: 859–62.

STARCK E, PAULOUCCI V, HERZER M, CRUMMY A (1984). Esophageal stenosis: Treatment with balloon catheters. *Radiology* **153:** 637–40.

THOMAS RJ (1981). The repsonse of patients with fistulas of the gastrointestinal tract to parenteral nutrition. *Surgery, Gynecology and Obstetrics* **153:** 77–80.

WILDER WM, MELHERN RE (1989). Balloon dilatation of post-surgical and ano-rectal strictures in two infants. *Pediatric Radiology* **19:** 527–9.

4
ANGIOGRAPHY

CHAPTER 11

Embolization of the liver, pancreas and gastrointestinal tract

Götz Richter and Gunter W. Kauffmann

Angiographic technique 176

Indications and contraindications 176

Pre-embolization patient preparation 178

Embolization materials and techniques 178

Complications, common mistakes and precautions 183

Results 184

References 188

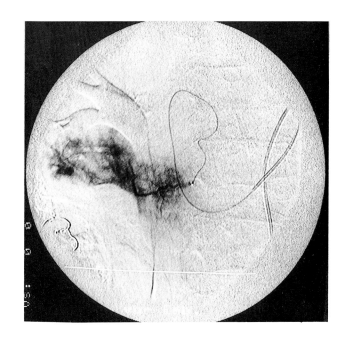

Embolization or, more precisely, transcatheter vascular occlusion therapy of visceral structures makes a significant contribution to the treatment of abdominal inflammatory disease, blunt or iatrogenic abdominal trauma and both benign and malignant tumours. It may have a curative role in a variety of lesions such as intractable bleeding or vascular malformations or may have a palliative role in the alleviation of symptoms of malignant disease.

ANGIOGRAPHIC TECHNIQUE

Coeliac axis and mesenteric arteriography can usually be carried out using a cobra or sidewinder catheter inserted via the femoral artery. However, if superselective catheterization is required for embolotherapy sophisticated catheter exchange techniques and the use of a variety of steerable guidewires may become necessary. In some instances (approximately in 10%) a brachial artery approach is used, particularly in very slender patients with a narrow angle between the coeliac axis and the aorta. The angiographer undertaking embolization procedures should be familiar with the many variants of visceral vascular anatomy.

Any visceral artery intervention requires a complete nonselective study of both the coeliac axis and the superior mesenteric artery. The next step is determined by the disease pattern to be treated. The mechanism of action of embolotherapy is determined by the primary level of vascular occlusion (Kauffmann and Richter, 1990). Hence, we distinguish between central, peripheral and capillary types of occlusion all of which have their special indications and technical requirements (Kauffmann and Richter, 1989, 1990). Some lesions need to be treated with the catheter tip positioned as superselectively as possible; others should be treated more centrally. There is no universally applicable method of embolization; a carefully selected and disease-adapted approach is required.

INDICATIONS AND CONTRAINDICATIONS

Liver embolization

Embolization is indicated in a variety of malignant and benign conditions. In some of these the indications are firmly established whereas in others embolization is carried out on an experimental basis (Clouse, 1989). None of the indications listed below are absolute; the advantages and disadvantages of embolization should be compared with those of other available methods of treatment. Pre-existing occlusion of the portal vein is a contraindication to the procedure. In the presence of an occluded portal vein arterial embolization can only be performed on a subsegmental level thus avoiding major parenchymal loss. Sepsis and decreased liver function due to any cause are also relative contraindications to embolization.

Malignant disease

Irresectable liver tumours, especially hepatocellular carcinoma (HCC), represent the most frequent indication for liver embolization (Clouse, 1989; Pelletier et al., 1990). In our department more than 90% of all gastrointestinal embolizations are performed for this indication. The primary purpose of the procedure is the palliation of symptoms such as abdominal pain and ascites. In some patients with localized, irresectable tumours embolization may prolong life expectancy. The approach to embolization in patients with HCC is influenced by the histology of the tumour. Three main types are identified: Expansive HCC usually has a tumour capsule and does not infiltrate portal or hepatic veins; it usually responds well to therapeutic embolization. The infiltrating type of HCC tends to invade vascular structures, is less localized and less amenable to embolotherapy. The diffuse type of HCC invariably infiltrates vascular structures and is usually widespread upon primary diagnosis. Patients with this tumour rarely benecit from therapeutic embolization.

In metastatic liver disease, therapeutic embolization is mainly carried out to relieve symptoms caused by the intrahepatic mass. Chemoembolization may lead to prolongation of life expectancy in some patients but oncological opinion on this subject remains divided (Schlag, 1991). The effectiveness of therapeutic embolization partly depends on the histological characteristics of the tumour. Colorectal primary neoplasms are the most frequent souce of liver metastases; such deposits usually cause no symptoms and therapeutic embolization is unnecessary in most patients. In contrast to these tumours, hypervascular metastases from malignant endocrine pancreatic or other intestinal tumours (apudomas) may produce clinical symptoms early in the course of the disease. For example, the carcinoid syndrome is characterized by episodes of flushing, vomiting and diarrhoea. In addition to causing metabolic disorders, hypervascular liver metastases tend to be more painful than colorectal metastases and therapeutic embolization may provide useful palliation.

Benign disease

Haemorrhage is the most frequent indication for embolization in benign disease (Clouse, 1989). A source of bleeding must be identified before the procedure is carried out. In general, haemorrhage from blunt abdominal trauma is managed either conservatively or surgically in the first instance. The main indication for embolization is the identification of a circumscribed intrahepatic vascular lesion, for example, a pseudoaneurysm, not accessible to surgery and causing measurable blood loss.

Arteriovenous malformations are rare. They should be embolized if they cause pain or if they lead to right heart failure.

Embolization is occasionally indicated in benign liver tumours. Ruptured adenomas have been embolized to control a life-threatening situation. Embolization has also been carried out in giant hemangiomas to reduce the volume of the lesion and the risk of haemorrhage.

Embolization of the pancreas

In general, the indications for embolization of the pancreas are less frequent and are less well defined than those for liver embolization. The pancreas is supplied by a network of vessels arising from the gastroduodenal artery (pancreatic head), the splenic artery (pancreatic tail) and the superior mesenteric artery (head, body). This multiple arterial supply may reduce the effectiveness of therapeutic embolization.

Malignant disease

In primary tumours of the pancreas there is almost no place for embolotherapy. In rare cases of recurrent hypervascular tumours like malignant apudomas (carcinoid tumours, gastrinomas etc.) superselective embolization may have a role in the management of intractable pain or other severe tumour-related symptoms.

Benign disease

The main use of embolization in benign disease is to control haemorrhage. Pancreatic haemorrhage typically results from pseudaneurysms caused by chronic pancreatitis. Such aneurysms usually arise in the splenic or gastroduodenal arteries. In the rare event of rupture therapeutic embolization represents a more rapid and efficient method of treatment than surgery.

Gastrointestinal embolization

The indications for gastrointestinal embolotherapy depend on the part of the bowel affected by the disease being treated and on the feasibility of surgical or endoscopic methods of management of the condition. Hence, embolization procedures within the arterial supply of the stomach or rectum are rare while embolotherapy is performed more frequently for duodenal patholoy (Rahn et al., 1982; Lang et al., 1990, Lang, 1992). Embolization is carried out rarely in the stomach and rectum partly because of the effectiveness of surgery or endoscopy in these organs but also because embolization may cause bowel gangrene if performed inappropriately. Capillary embolization should be avoided because of the risk of gangrene. However, delivery of emboli into large, proximal vessels is ineffective. This is an area in which attention to technical detail is vital for the achievement of satisfactory results.

A frequent problem is the difficulty of demonstrating the source of bleeding in many patients. Direct angiographic visualization of intestinal haemorrhage requires a substantial rate of blood loss at the time of the procedure. Unfortunately, by the time most patients are brought to the angiographic table the bleeding has stopped.

In patients with gastric haemorrhage embolization is only indicated if both endoscopy and surgery in combination with an aggressive haemostatic drug regimen have already failed to control active bleeding. Indications include inoperable tumours, bleeding caused by chemotherapeutic agents (particularly in lymphoma), anastomotic bleeding after gastric surgery (particualry in septic patients) and ruptured pseudoaneurysms.

Duodenal embolization, too, is considered only as second-line therapy after the failure of surgery or endoscopy. The commonest source of duodenal haemorrhage is a pseudoaneurysm of either the gastroduodenal artery or one of its major branches such as the superior or inferior pancreatoduodenal arteries and the right gastroepiploic artery. Pseudoaneuryms of these vessels may develop as a result of atherosclerotic disease or as a late and rare complication of pancreatitis. Duodenal haemorrhage may also occur following surgery such as Billroth I gastrectomy, or following papillotomy.

Small and large bowel embolization are rare procedures performed in carefully selected cases. In contrast to liver, gastric or duodenal artery embolization there is a significant risk of infarction because of the end-artery character of the vascular supply to the bowel. If satisfactory results and a low complication rate are to be achieved it is essential to employ a careful embolization technique based on detailed knowledge of the arterial supply to the bowel. The five bowel embolization procedures performed in our institution during the last six years included one case of intestinal telangiectasia associated with severe gastrointestinal haemorrhage and portal hyperternsion, one case of lymphoma causing bleeding during chemotherapy, a case of haemorrhage arising from a caecal vascular malformation in a patient who refused surgery, and postanastomotic haemorrhage in two patients who had undergone rectosigmoid resection.

Pre-embolization patient preparation

Therapeutic embolization should be performed under close monitoring of the patient's pulse rate, blood pressure and peripheral oxygen saturation. If a long embolization session is anticipated it may be useful to institute transcather bladder drainage before the procedure.

When considering liver embolization the adequacy of liver function and patency of the portal vein should be established prior to the procedure. In cases of embolization for the treatment of liver tumours it is important to identify any arterial feeders and arteriovenous or arterioportal shunts. Pre-embolization imaging should be used to define tumour volume and intrahepatic spread because post-procedural restaging and assessment of the success of the procedure may be difficult without baseline studies.

In pancreatic embolization there is no need for special preparation other than adequate coeliac an-

giography and delineation of the underlying disease.

In gastric, duodenal and mesenteric embolization haemorrhage represents the most frequent indication for therapeutic embolization. Endoscopy prior to the procedure may demonstrate the site of bleeding. In upper as well as in lower gastrointestinal haemorrhage preprocedural definition of the source of bleeding helps to shorten significantly the length of the procedure. Other preparation steps depend on the condition of the patient at presentation. Patients in haemodynamic shock should be stabilized prior to the procedure. The ability to control breathing and motion is crucial for the initial diagnostic angiographic studies as well as for the procedure itself and general anaesthesia may be necessary for this purpose. In patients with haemorrhage coagulation abnormalities, either pre-existing or caused by the clinical situation, may diminish the effectiveness of embolization. Such patients may require transfusion of fresh frozen plasma to correct the coagulation abnormality. In addition, the embolization technique used in such patients should not rely exclusively on resorbable material.

Embolization materials and techniques

Three basic considerations should be taken into account when selecting an appropriate embolization material:

1 No single material is suitable for all purposes and in each clinical situation certain substances are more appropriate than others.
2 The interventionist should understand the mechanism of action of the various substances and the most appropriate delivery technique in each case.
3 It is wise to concentrate on the use of a few substances to build up sufficient individual experience with each material.

Liver embolization

Pseudoaneuryms should be treated as selectively as possible, preferably, by directing the material into the lesion itself (Coldwell, 1990). In our opinion the material of choice is the occlusion (micro) coil. For the majority of cases a coaxial catheter technique is required. The choice of guiding catheter depends on the anatomy of the coeliac trunk. The tip of the embolization catheter should be advanced at least into the common hepatic artery and if possible into the right or left hepatic artery. By the aid of road-mapping techniques or high-quality on-screen storage of a diagnostic run carried out via the guiding catheter, a microembolization catheter such as the steerable Tracker 18 system (Target Therapeutics) is guided more peripherally towards the lesion. The catheter can, usually, reach the segmental arteries and their major branches. Certain technical modifications help to improve the versatility of the system such as the use of a Y-connector between the Tracker 18 catheter and the guiding catheter and a mild J-curve at the tip of the wire that comes with the system. The catheter tip should ideally be positioned within the pseudoaneurysm itself; if this is not possible it should be placed within the arterial feeder of the lesion. A sufficient number of Tracker microcoils are delivered into the lesion closely following the instructions that come with the coil kit (coils, coil feeder and special pusher wire). The choice of coil sizes (2–4 mm) depends on the size of the lesion. It is important to note, however, that the occlusive property of the Tracker micro coils is not as great as that of large size coils (0.035 inch wire). Hence, in some cases the embolization may be completed by pushing gelatin sponge particles into the lesion once the coils are in place. Such particles should be cut from a sponge block as stripes with a length of 4–8 mm and a diameter of 1–3 mm and backloaded into a 1 ml tuberculine syringe with the aid of which they can be injected into the target. The use of the Y-connector allows demonstration of success without changing the catheter position.

Embolization of benign liver tumours causing haemorrhage or other symptoms also requires a subsegmental approach and a coaxial catheter technique. The choice of material may vary from lesion to lesion. In giant haemangioma or bleeding liver adenoma without evidence of arteriovenous shunting, peripheral embolization with small to moderate amounts of Ethibloc (Ethicon, Norderstedt) or non-resorbable particulate matter may be most appropriate. In arteriovenous malformations a different approach is necessary: while embolization of such lesions should be performed on a subsegmental level and as close to the lesion as possible, special attention must be paid to arteriovenous shunts. The safest and quickest technique is to deploy coils within the feeder artery, as many as necessary to form a functioning matrix for sponge particles to lodge within the coil framework when injected thereafter. This prevents emboli reaching the pulmonary circulation via arteriovenous shunts. In extremely high flow lesions additional central occlusion of the right or left

hepatic artery may help to avoid recanalization. It should be noted, however, that central occlusion alone is absolutely worthless in this type of lesion.

A variety of embolization substances and techniques has been proposed over the years for the embolization of primary and secondary malignant liver tumours. The choice of material and technique depend on the type of lesion. There are three basic types of HCC: the expanding type (unilocolar or multiocular), the infiltrating type and the diffuse type, all of which may be complicated by the presence of cirrhosis. Diffuse HCC is almost invariably associated with early onset of liver dysfunction which is a contraindication to embolization. In the expanding type angiography typically depicts a relatively well-defined lesion arising from and fed by a distinct segmental liver artery (unilocular) or by several such vessels (multilocular). Our approach here is as follows: the tip of the embolization catheter is position within the tumour feeding artery. Then highly selective *cis*-platinum perfusion of the tumour is performed with a dosage equal to 50% of the total dose appropriate for a single intravenous treatment cycle (Fig. 11.1 and 11.2). In many cases a similar amount is used to prepare a suspension with 20–40 ml Lipiodol which is then injected into the lesion. To complete the procedure a small amount of Ethibloc (Ethicon, Norderstedt, Germany) is injected to block the feeding artery. This technique is applicable to any tumour nodule. In the infiltrating tumour type, usually subsegmental or even segmental embolization is impossible because of trans-segmental tumour spread. The technique we employ in such cases is as follows: before embolization is performed a computed tomographic (CT) arteriography study is carried out, with the catheter tip already in a position suitable for embolization. This helps to define whether all tumour tissue will be reached or whether redirection of the catheter tip is necessary. Computed tomographic arteriography is followed by chemoperfusion and chemoembolization with *cis*-platinum, often combined with Lipiodol, as described above. Ethibloc embolization, however, is avoided. Within a period of 24–72 hours a CT study is done to establish the extent of tumour embolization, which is reflected in the Lipiodol distribution within and around the tumour. A CT study performed one month after the procedure demonstrates the effectiveness of embolization. If the tumour response is unsatisfactory the procedure is repeated. Primary occlusion of the hepatic artery proper should be avoided in order to maintain access for any further embolization procedures, should these prove necessary.

In patients with metastatic deposits the approach

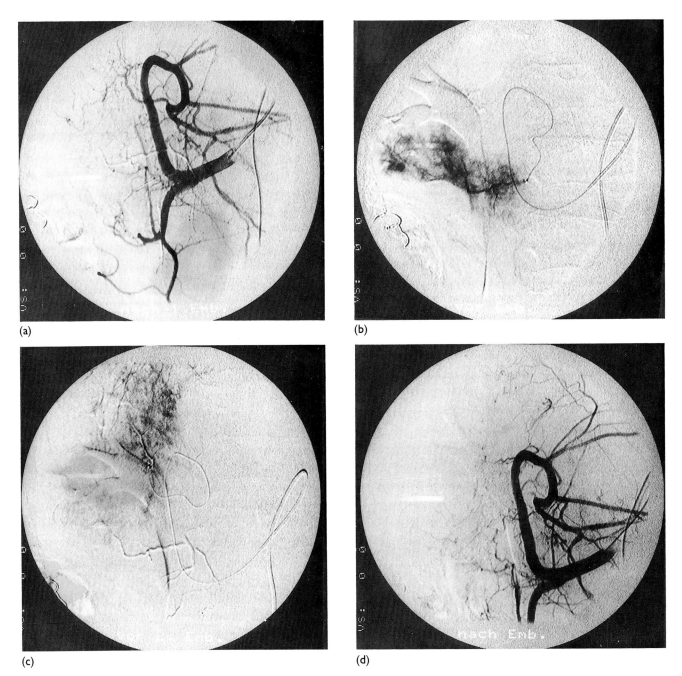

(a)

(b)

(c)

(d)

Fig. 11.1

Superselective chemoembolization of recurrent hepatocellular carcinoma 14 months after right hemihepatectomy. The superselective approach was chosen because segments II and III seem to be completely free of tumour involvement. (a) Common hepatic arteriogram using a sidewinder 3 catheter (5 Fr): the right branch of the proper hepatic artery was ligated during surgery. Two small branches arising from the left hepatic artery supply the recurrent tumour. The main side branches (segmental arteries to segments II and III) are normally perfused. (b) A Tracker 18 catheter is manipulated into a small side-branch in the caudal part of the recurrent tumour, Cis-platinum of 25 mg, followed by injection of 1.5 ml Ethibloc were administered. (c) Another Tracker 18 catheter is manipulated into the artery supplying the tumour more proximally to perform angiography and superselective chemoembolization. Cis-platinum, 25 mg, followed by 1.5 ml Ethibloc were administered. The faint grey shadow is due to the previous Ethibloc injection. (d) Arteriography after chemoembolization: tumour vasculature is no longer visible. Comment: This patient survived for two years, following the procedure; he died from multiple pulmonary metastases.

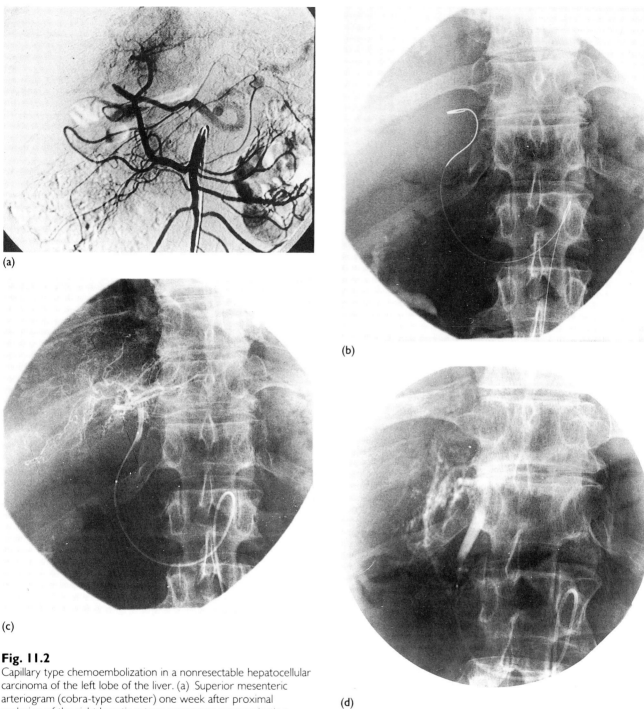

(a)

(b)

(c)

(d)

Fig. 11.2
Capillary type chemoembolization in a nonresectable hepatocellular
carcinoma of the left lobe of the liver. (a) Superior mesenteric
arteriogram (cobra-type catheter) one week after proximal
occlusion of the right hepatic artery: numerous communicating
vessels are seen between the superior mesenteric artery and the gastroduodenal artery. The left hepatic artery supplies a hypervascular
tumour mass; the right hepatic artery is occluded proximally. (b) Superselective advancement of a highly flexible but stiff-bodied guidewire
with a 1:1 torque response (0.020 inch, gold-tip 6 cm long, Schneider S.A., Lausanne, Switzerland). The stiff part reaches the hepatic artery
bifurcation. (c) A custom-made occlusion balloon catheter (5 Fr shaft, 4 mm diameter and 1.5 cm length of balloon) is advanced into the left
branch of the hepatic artery proper for occlusion angiography which demonstrates extensive neoplastic circulation within the left lobe.
(d) Magnified view during injection of Ethibloc assisted by balloon occlusion to avoid reflux. Cis-platinum 55 mg, had already been injected.
(e) A computed tomographic study one week after chemoembolization demonstrates widespread tumour necrosis. Some white parenchymal
staining and opacification of medium sized arterial vessels from Ethibloc is still visible. (f) Same study four sections inferior to the one in (e).
Comment: Arterial redistribution embolization and occlusion-assisted Ethibloc injection were done to avoid reflux and inadvertent healthy
tissue damage while allowing as much embolic material as possible to reach the tumour. The patient survived 17 months and died from a
second tumour which developed in segment VI.

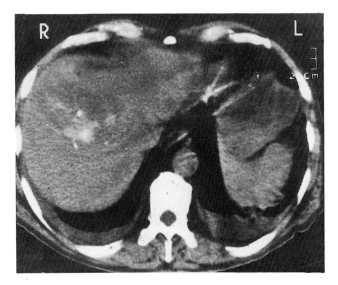 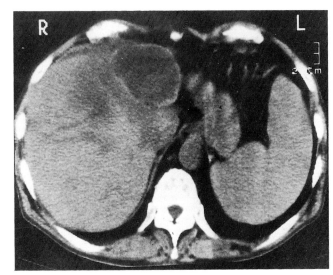

Fig. 11.2 *continued*

to embolization is slightly different to that adopted in HCC. The metastases most suitable for embolization are hypervascular lesions such as those arising from neuroendocrine tumours or renal carcinoma. The main purpose of the procedure is palliative and it would be illogical to try to destroy a specific deposit. The hypervascularity of the lesions is used to flow-direct the embolization material from a central catheter position. The tip of the embolization catheter is placed either within the hepatic artery proper or within the right or left hepatic arteries. We prefer to use a coaxial system which allows better steerability and control during the procedure and also makes catheter exchanges considerably easier. Nonresorbable small sized particulate embolization material (e.g. polyvinyl alcohol particles, 40–120 τ size) is injected until peripheral arterial stasis begins. This is followed by injection of small amounts of Ethibloc which solidifies thus becoming an intra-arterial plug which helps to prevent recanalization.

Pancreas embolization

The main indication for embolization of the pancreas is a bleeding pseudoaneurysm. Embolization should be preceded by detailed diagnostic angiography, including superselective studies of the splenic, gastroduodenal and superior mesenteric arteries, in order to demonstrate the vessels feeding the false aneurysm. As the pancreas is fed by a network of vessels a false aneurysm usually has a 'front door' and a 'back door'. In view of this, the embolization technique should aim to achieve direct occlusion of

the aneurysm. Usually, this can only be achieved by a coaxial catheter technique using highly steerable small sized systems such as the Tracker 18 catheter. The aneurysm should be approached via the biggest feeder artery. A 5 Fr guiding catheter should be advanced as close to the lesion as possible. This manoeuvre may be aided by intra-arterial antispasmodic medication (e.g. 100 μg nitroglycerine). Then, the Tracker 18 catheter should be advanced further towards the lesion, preferably into the aneurysm. Only this guarantees a complete cut-off of blood supply from antegrade and retrograde collateral vessels. The occlusion material of choice is micro coils (Tracker micro platinum coils) in a size appropriate for the size of the lesion and the feeding artery. If direct occlusion of the aneurysm proves impossible, even using a coaxial catheter technique, the embolization has to be completed in two steps: first the antegrade feeder artery ('front door') is occluded followed by occlusion of the retrograde feeder artery ('back door'). In such cases several diagnostic superselective angiography runs are necessary for full demonstration of the relevant anatomy. In patients with abnormal coagulation in whom direct embolization of the aneurysm cannot be achieved even the above method may fail to control bleeding. If this is the case, and particularly if angiography still shows some flow to the aneurysm, a fine Ethibloc bolus (0.1–0.3 ml) may help to occlude the arteries. It has to be noted, however, that great care is necessary to avoid overembolization because this may lead to organ necrosis. Ethibloc is also a suitable material for embolization of recurrent pancreatic apudoma recurring postoperatively. Ethibloc is injected into the tumour via the biggest feeder artery, using a coaxial technique and taking care to avoid reflux into major vessels.

Gastrointestinal embolization

The basic technical principles of pancreatic embolization – superselectivity, the use of a coaxial technique and a direct approach to the lesion – also hold true for the vast majority of bowel lesions which need to be embolized. The main differences are in the choice of guiding catheters. Embolization of the left gastric artery is often more technically demanding than embolization of other visceral arteries. A useful manoeuvre is the insertion of a sidewinder III some distance into the coeliac trunk followed by withdrawal and anterior rotation. In other cases a sidewinder I proves more useful, and sometimes a brachial approach is necessary. Access to the gastroduodenal artery is determined primarily by the anatomy of the coeliac trunk. In approximately 50% of patients a cobra-type catheter (5 Fr) will easily enter the gastroduodenal artery. In other patients, initial use of a sidewinder 2 or 3 catheter is necessary, followed by a later exchange for a cobra catheter. The superior and inferior mesenteric arteries can be catheterized with cobra or sidewinder catheters. In cachetic patients the use of sidewinder catheters is more frequently successful.

When performing gastric or duodenal embolization we use the same embolic materials (primarily coils, plus a very small amount of Ethibloc) as described above for pancreatic embolization.

Mesenteric artery embolization requires great caution if bowel necrosis is to be avoided. The embolization coil, particularly the micro coil, is the embolic material of choice. It should be placed as close to the lesion as possible. Underembolization, allowing some flow towards the lesion, is more acceptable than in other regions. Mere reduction of blood flow by the insertion of a few coils combined with correction of any coagulation abnormality may be clinically sufficient. In our opinion there is only one exception to this approach which is embolotherapy of mesenteric telangiectasia. In such cases a combination of multiple coils lodged within the arterial feeders of the malformation, followed by complete obstruction of these vessels by Ethibloc, may be the only way to decrease portal hypertension and variceal or mucosal bleeding.

COMPLICATIONS COMMON MISTAKES AND PRECAUTIONS

Liver embolization

The simplest complication is vascular damage caused by catheter introduction; this may cause severe vasospasm, intimal tears or dissection. Such problems have been decreased significantly by the use of coaxial systems. It is wise to use such a system early on during the procedure thus avoiding forceful peripheral advancement of the guiding catheter. However, the use of coaxial catheters is not free of complications, some of which are specific to the system being used: thromboembolic complication or adherence to the inner catheter to the outer catheter due to thrombus may develop; this can be prevented by using a Y-connector and constant or intermittent flushing of the guiding catheter. Reflux and inadverdent transport of embolic material to inappropriate areas is a dangerous complication which may occur with any embolic substance used. If this happens with coils in the liver there are usually no significant clinical sequelae. Even coil dislodgement in the gastroduodenal artery may have no unfavourable sequelae because of the presence of a good collateral supply. Nevertheless, inadvertent passage of a coil through a side-hole may cause a severe problem. The coil may leave the catheter only partially and when a retrieval manoeuvre is tried it may enter a critical vascular area. In view of this potential problem coils should be deployed through single end-hole catheters only. When liquid embolization substances are used the danger from reflux is much more significant, particularly in the case of the cystic and gastroduodenal arteries. Embolic material may enter the cystic artery during embolization or even chemoperfusion of HCC from a very proximal catheter position. The risk is even greater in the case of the gastroduodenal artery. In view of this, whenever substances toxic to mucosal tissue have to be injected into the liver the cather tip needs to be beyond the origin of the above-mentioned arteries. In cases where such precautions are impossible coil occlusion of these arteries to redirect flow is necessary. The presence of arteriovenous shunts is associated with a risk of pulmonary embolic complications upon embolization. In such cases the use of small sized particulate materials cannot be recommended. The embolization technique will depend on the shunt size. In very large shunts occlusion balloon catheters are helpful to decrease antegrade flow while still allowing the passage of a small calibre Tracker

18 catheter. Balloon occlusion catheters inserted via a transjugular or tranfemoral venous approach can also be used to block the venous outflow thus preventing upstream migration of particular emboli.

The so-called postembolization syndrome (fever, nausea, vomiting, headache, abdominal pain) is related to the extent of tumour necrosis after embolization. Pain can be a significant problem when larger amounts of Ethibloc are used. The pain can be attributed to its alcohol content. In chemoembolization postprocedural vomiting and cardiovascular depression are frequent side-effects of the anticancer drugs (e.g. *cis*-platinum).

Pancreas embolization

The catheter- and material-related problems are virtually identical to those in liver embolization. When using large particles such as coils for embolization organ-specific complications are relatively infrequent. One area in which special difficulties arise is the pancreatic head when the catheter approach is via the mesenteric artery. In such cases, coil misplacement or dislodgement into the main stem of the superior mesenteric artery may result in fatal bowel necrosis. When liquid emboli are used this danger is even greater. The use of liquid substances also carries a risk of organ necrosis from overembolization. In view of these considerations alcohol is best avoided in embolization of the pancreas.

Gastrointestinal embolization

The main risk of embolization of gastrointestinal lesions is bowel necrosis due to occlusion of the superior or inferior mesenteric or their branches. Very distal embolization is also dangerous as there may be insufficient collateral circulation to prevent gangrene. Liquid emboli must be reserved for exceptional cases and very special situations (abnormal coagulation, extreme clinical urgency, coils already in place to avoid capillary transport). Only experienced interventionists should consider using liquid substances in this area; Ethibloc is a safer and thus preferable embolic material.

RESULTS

Liver embolization

The success rate of therapeutic embolization of benign liver lesions is high. Since micro embolization catheters became available approximately five years ago we have not failed to embolize successfully a hepatic artery pseudoaneurysm. A total of 23 such pseudoaneurysms have been embolized, most by using the Tracker 18 system in combination with micro coils. Two pseudoaneurysms resulting from vascular injury during percutaneous biliary drainage were occluded successfully. Two patients with severe gastrointestinal telangiectasia causing chronic right heart failure were treated by repeat arterial occlusion starting at a very peripheral level (Fig. 11.3). A combination of gelatin sponge particles and large and medium sized coils achieved substantial reduction of the arteriovenous shunt flow while avoiding hepatic parenchymal necrosis or liver failure. However, in both patients multiple embolization sessions (six in one case and eight in the other), were necessary. There were no complications from the procedure in any of the patients with benign liver disease. In the majority of our patients with hepatic metastases embolization was highly effective and relieved severe clinical symptoms. From 1988 to 1992 we treated nine patients with hypervascular liver metastases from endocrine tumours. In six patients a total or near total tumour necrosis was seen during follow-up. In the other three only mild reduction in size of the metastatic deposits was seen on CT and ultrasound scans. Four patients with a giant single metastasis from colorectal cancer who were not fit for surgery were managed using a combination of high-dose chemoperfusion (mitomycin C and *cis*-platinum) and segmental Ethibloc embolization. In all four patients widespread necrosis was confirmed by CT. The average survival time of the patients is 23 months. One patient had died from multiple lung metastases and in a second one there is regrowth of the metastasis after a 25 month period of regression. In the other two patients the tumours appear to be under control during follow-up.

In 22 of 31 patients with expanding HCC managed using a combination of chemoperfusion and segmental chemoembolization there was total tumour necrosis. In the other nine patients substantial but incomplete tumour necrosis was seen on CT scans. In these patients tumour progression was found during follow-up. The average survival time of this subgroup of patients is 19 months whereas in patients with complete initial tumour response it is

25 months. In the latter group life expectancy is determined by metachronous tumour development. This was seen in seven of the initial group of 31 patients. Embolization of the infiltrating type of HCC usually fails to achieve total tumour control but may achieve satisfactory palliation. The average survival time of 19 patients with this type of tumour was 11 months. The corresponding figure in similar patients not referred for treatment is 4–5 months. However, it should be noted, that the longer life expectancy is achieved at the expense of repeat hospitalization as an average two to three sessions of treatment is necessary.

Although in general the results achieved in our unit when treating malignant liver disease with embolization have been good, it should be emphasized that the group of patients treated was highly selected. Nevertheless, selection is necessary if this form of treatment is to be used most appropriately.

Pancreatic embolization

In this organ a 100% technical success rate was achieved when embolizing pseudoaneuryms (11/11 patients). However, the clinical success rate was largely determined by the clinical state of the patient at presentation. This is largely determined by the amount of retroperitoneal or intra-abdominal blood loss. Thus the 30 day mortality was 18% despite technically successful and uneventful embolization. Patients may die when the time gap between the onset of symptoms and treatment is too long and multiorgan failure develops.

Two patients were embolized for local recurrence of endocrine pancreatic tumours using a coaxial technique for the injection of Ethibloc. In both some tumour regression was observed. Computed tomography demonstrated tumour regrowth 14 and 17 months, respectively, after initial embolization. Angiography demonstrated vascular tumour supply arising from fine retroperitoneal collaterals not accessible for embolization.

Gastrointestinal embolization

The technical and clinical success rates of gastrointestinal embolizations are somewhat lower than those achieved in other vascular territories. Pancreaticoduodenal lesions may have so many tributaries from the coeliac axis that only direct embolization of the lesion itself guarantees success. Since micro embolization catheter systems became avail-

able in 1988 we have treated 24 patients with vascular lesions in the area of the duodenum and the pancreatic head. In 15 patients we were unable to achieve complete occlusion of the pseudoaneurysm or other vascular lesion. Such cases were managed by embolization of the major feeders (Fig. 11.4). In six patients in which a combination of liquid and particulate emboli was ussed haemostasis was achieved. However, mucosal necrosis was seen in two patients, necessitating emergency surgery. In the other nine patients the major feeding artery was occluded with coils and gelfoam particles until opacification of the lesions was no longer seen. Nevertheless, in five of these patients a second session was necessary to occlude another feeding artery not seen during the initial procedure. The technical success rate (complete occlusion of the lesion, no recanalization during the 30 day post-treatment period) in this group of 24 patients was 100%. This compares favourably with our results in the early 1980s when success rates did not exceed 75%. Failures usually resulted from inability to access the lesion or even the main feeders because micro catheter systems were not available. The clinical results are relatively unsatisfactory because of the poor state of most patients at the time of presentation. The 30 day survival rate in the abovementioned group of 24 patients was 87% (21/24). Three patients died from multiorgan failure. The complication rate was 17% (4/24): two patients developed duodenal necrosis, one pancreatitis and another permanent renal failure.

Mesenteric artery embolization (colon and rectum) is performed very infrequently. The treatment of choice for most colonic, sigmoid or rectal lesions is either surgical or endoscopic. During a period of 15 years in two different institutions (departments of radiology of the university hospitals in Freiburg and Heidelberg) only 11 embolizations in this vascular territory have been performed. The patients treated included some with haemorrhage caused by intestinal lymphoma or by the chemotherapeutic agents used to treat this condition. Other patients were embolized because surgery was refused for religious reasons or because the patients were unfit for anaesthesia. The relatively high re-intervention rate of 64% (7/11) reflects the fact that we tried to use the minimum amount of embolic material necessary to achieve haemostasis and avoided distal embolization dose to the bowel wall. This policy allowed us to achieve a high technical success rate with only one case of bowel necrosis. One patient developed an ischaemic stricture after embolization of lymphoma of the ascending colon.

Gastrointestinal embolization should never be undertaken without careful consideration of the clinical

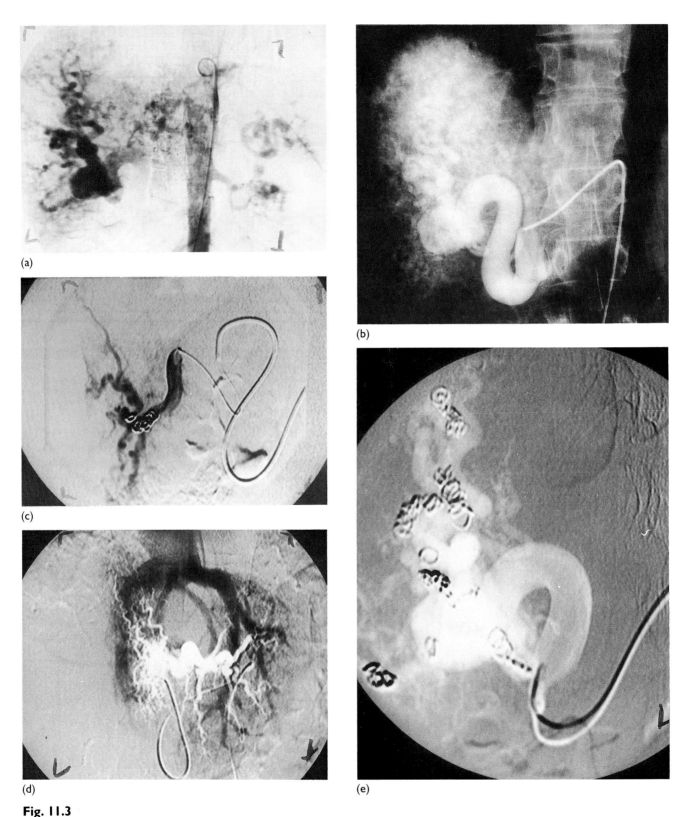

Fig. 11.3
Telangiectasia of the liver: a 55 year-old patient presenting with right heart failure caused by huge arteriovenous malformations throughout the liver. (a) Aortography demonstrates extensive malformations of the intrahepatic arteries. The main stem of the common hepatic artery is grossly enlarged. (b) Common hepatic arteriogram shows widening and elongation of the common hepatic artery and its right and left branches. All intrahepatic segmental and subsegmental arteries are markedly enlarged. (c) Coaxial embolization: a cobra catheter has been advanced distally into the right hepatic artery and a Tracker 18 catheter has been manipulated into the artery segment V. Despite deployment of seven micro coils into subsegmental branches peripheral flow is still present. In view of this, large-size gelfoam particles were injected

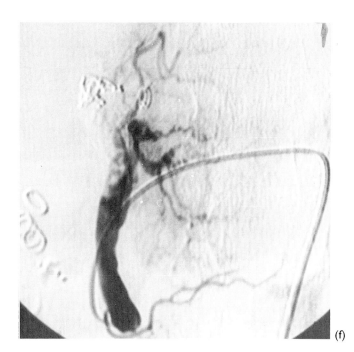

(f)

subsequently until flow stopped. (d) One week later: radiograph taken during coil embolization of segments VI, VII and VIII. At the time of this frame roughly 30 coils ranging in sizes from 5 to 7 mm had been deployed. Because flow was still present several gelfoam particles were injected until contrast material was beginning to reflux out of the embolized vessel. (e) Another week later: left hepatic arteriogram shows massive arteriovenous shunting; (middle and left hepatic veins in black; arterial phase of segments I, II and III in white). (f) Angiography after deployment of eight coils (6 and 7 mm) and injection of approximately 30–40 gelfoam particles (2 × 2 mm size) through a cobra catheter superselectively positioned within the left hepatic artery. Peripheral staining or arteriovenous shunting are not longer seen. Comment: The procedure was carried out in a staged fashion over a one month period to avoid sudden liver failure and to allow portal flow compensation. A combined peripheral and proximal approach is mandatory in such cases. Central artery occlusion would allow collaterals to grow leading to recurrence of arteriovenous shunting. On the other hand, small particles should be avoided as they could pass through the shunt. Isotope scans one week after completion of treatment showed dramatically decreased left-to-right shunting.

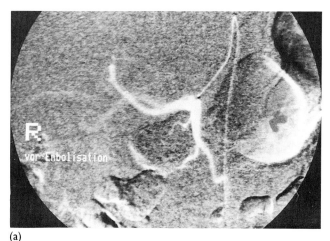

(a)

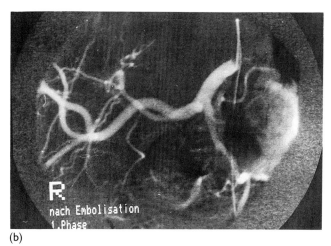

(b)

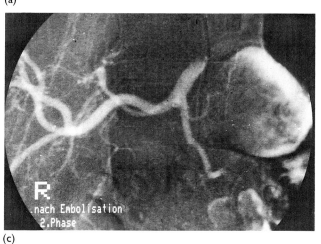

(c)

Fig. 11.4

Severe upper gastrointestinal bleeding from a duodenal ulcer eroding the main stem of the gastroduodenal artery. Superselective embolization of the bleeding site itself and the distal portion of the gastroduodenal artery. (a) Contrast medium injection after catheterization of the common hepatic artery with a sidewinder 1 catheter through which a Tracker 18 catheter has been advanced near the office of the gastroduodenal artery. Massive contrast extravasation from the main stem of the inferior pancreatoduodenal artery into the duodenal lumen is demonstrated. Following this arteriogram the Tracker catheter was advanced into the inferior pancreatoduodenal artery and 8–10 medium sized gelfoam particles (1 × 1 mm as dry cubes) deployed. (b) Arteriography following superselective embolization of the inferior pancreatoduodenal artery; the Tracker catheter has been withdrawn. There is now no evidence of extravasation from the inferior pancreatoduodenal artery itself although a small amount of contrast medium is extravasating from a more distal side-branch. In view of this a Tracker 18 catheter was advanced into the distal gastroduodenal artery beyond the possible source of the extravasation and another 5–7 medium sized gelfoam particles were deployed. (c) Angiogram after completion of embolization: no more contrast extravasation was seen at either of the previous bleeding sites. Comment: the patient's condition improved significantly following embolization. He was discharged from hospital one week later.

state of the patient and detailed discussion with the clinical teams involved in his or her care.

REFERENCES

CLOUSE MC (1989). Hepatic artery embolization for bleeding and tumours. *Surgical Clinics of North America* **69**: 419–32.

COLDWELL DM (1990). Hepatic arterial embolization utilizing a coaxial catheter system: technical note. *Cardiovascular and Interventional Radiology* **13**: 53–4.

KAUFFMANN GW, RICHTER GM (1989). Embolisation der Niere. In: Interventionelle Radiologie. R. Günther, Th. Thelen eds. Thieme, Stuttgart.

KAUFFMANN GW, RICHTER GM (1989). Perkutane Embolisation/Okklusion – Embolisationsmaterialien. In: Interventionelle Radiologie. R. Günther, Th. Thelen eds. Thieme, Stuttgart.

KAUFFMANN GW, RICHTER GM (1990). Angiographic management of malignant tumors in the thorax, abdomen and bones – renal tumours: experimental results. In: Interventional Radiology. R. F. Dondelinger, P. Rossi, J. C. Kurdziel, S. Wallace eds. Thieme, Stuttgart, New York.

LANG EV, PICUS D, MARX MV, HICKS ME (1990). Massive arterial hemorrhage from the stomach and the lower oesophagus: impact of embolotherapy on survival. *Radiology* **177**: 249–52.

LANG EK (1992). Transcatheter embolization in management of hemorrhage from duodenal ulcer: Long-term results and complications. *Radiology* **182**: 703–7.

PELLETIER G, ROCHE A, INK O, ANCIAUX ML, DERHY S, ROUGIER P, LENOIR C, ATTALI P, ETIENNE JP (1990). A randomized trial of hepatic arterial chemoembolization in patients with unresectable hepatocellular carcinoma. *Journal of Hepatology* **11**: 181–4.

RAHN NH III, TISHLER JM, HAN SY, RUSSINOVICH NAE (1982). Diagnostic and interventional angiography in acute gastrointestinal bleeding. *Radiology* **143**: 161–6.

SCHLAG P. (1991). Surgical and adjuvant attempts to reduce liver metastasis in colon cancer. *Onkologie* **14**: 108–14.

Vascular stents in liver disease

G. M. Richter and J. C. Palmaz

Description of types of stent   190

Inferior vena cava stents   190

Portocaval stents   192

The rationale of stent choice for TIPSS   205

References   205

DESCRIPTION OF TYPES OF STENTS

There is a variety of vascular stents on the market which offer promise for the treatment of liver disease of vascular origin. There is a strong trend towards accepting that for each of the stents special indications are evolving or are already pertinent. It is our belief that this holds particularly true for stenting in liver disease.

Vascular stents are implanted as permanent devices. The choice of material and design properties, therefore, must be directed towards maximum intracorporeal durability. In this regard, factors such as metal corrosion, structural weak points or histological inertness are of the utmost importance. In the long history of medical applications stainless steel 316L has been shown to have excellent biocompatibility properties without any known corrosion problems. By contrast, tantalum alloys – the material used for the Strecker stent – are prone to relatively rapid intravascular corrosion. In addition to possible corrosion, specific design problems, or weak points, may add to the failure of long-term durability. Therefore, in the design of a permanent vessel implant crossing points with metal riding over metal ('fretting corrosion') should be reduced to a minimum.

Two different types of stent are widely used for the transjugular intrahepatic portosystemic stent shunts (TIPSS) the balloon expandable Palmaz iliac stent (Johnson and Johnson Interventional Systems, Warren, NJ) and the self-exapandable Wallstent (Schneider, Zurich, Switzerland). The Palmaz stent consists of a single segment tubular stainless steel mesh, surgical grade 316L with a wall thickness of 150 µm (= 0.004 inch) and an unexpanded length of 30 mm. Around its circumference, each stent has four staggered, offset slots that are 4.5 mm long. Its diameter in the non-expanded state measures 3.1 mm. Radial compression of the stent over a balloon catheter decreases this diameter by closure of the slots and, thus, firmly attaches the crimped stent to the catheter (7 Fr shaft). The Palmaz stent is a typical malleable stent type which, when expanded, employs plastic forces to withstand hoop stress. As it has no expansile force by itself, expansion of the coaxial balloon used for stent delivery will impose a force beyond the elastic limit of the metal mesh and, thus, radially open the slots to quadrangles. It is because of this mode of action that such a stent has a wide range of possible luminal diameters which are solely dependent on the size of the balloon used for delivery. The Palmaz stent is the only one which allows such a variability. The size of the balloon applied for stent deployment determines the final expansion ratio. By contrast, the Wallstent (Schneider) is a self-expandable vascular endoprosthesis, the design of which closely resembles a 'chinese finger'. The basic metallic material from which the stent is manufactured is a commercial secret. Because of the springload of the design a special delivery system is provided which consists of a 7 Fr catheter at the distal end of which the stent is both held in place and flattened down by a plastic membrane that covers the whole set. A side-arm is connected to the narrow interspace between the shaft and the membrane. As soon as contrast medium is injected at a limited pressure of 4–6 atmospheres through this side-arm, friction between membrane and catheter shaft is minimized and the membrane can be pulled back. This pulling or rolling back of the membrane will set free the stent from distal to proximal as it pops up from its springload. As the stent is very flexible it will open even when heavily bent or curved without kinking.

In addition to the two other types of stent previously mentioned the Gianturco stent (Cook) in the modified version by Rösch et al. (1987) has been advocated for stenting of hepatic veins or the inferior cava to treat other liver disease affecting vascular structures. This stent is also categorized as self-expanding and is made from springloaded stainless steel wire which is bent in a zigzag pattern and soldered at the ends to form a closed circle. The thickness of the wire, its inherent springload, the number of the turns and the angulation of the turns determine the final diameter and the resistance to hoop stress. Usually, a single stent cylinder is 2.5 cm in length. For long segment stenting double or triple cylinders are available. For deployment the stent can be pushed through a delivery cartridge in a compressed state. A specific feature of this stent is its availability in large diameters which are particularly useful in venous stenting.

INFERIOR VENA CAVA STENTS

Introduction

A variety of liver diseases may cause symptoms associated with obstruction of the hepatic part of the inferior vena cava such as primary and metastatic liver tumours the Budd–Chiari syndrome with or without caval webs, caudate lobe hypertrophy in liver cirrhosis, tumours of the caudate lobe, anastomotic strictures after liver transplantation and some rare tumours of the intrahepatic caval wall. In addition to symptoms specific for the causative disease, obstruction of the inferior vena cava above the inflow of renal veins induces significant pelvic and lower extremity venous congestion (inferior vena cava syndrome). In the end, this may lead to severe diminution of quality of life due to problems such as severe leg swelling, pelvic pain, bleeding upon defaecation and other symptoms within the pelvic and genitourinary organs attributable to venous congestion. Even a decrease of renal function may develop if collateral pathways fail to maintain sufficient venous renal blood flow.

Any treatment concept ideally has to target directly the underlying disease. Only if such treatment fails or if no specific treatment option exists are other treatment modalities appropriate; most of these are palliative. Therefore, vascular stenting of the inferior vena cava is practically the last option in the therapeutic armamentarium.

Among the diseases causing obstruction of the inferior vena cava the Budd–Chiari syndrome complex is quite unique. It may develop from suprahepatic caval abnormalities, intrahepatic venous lesions and pathologic changes affecting small veins. Furthermore, the rate of development of symptoms differs widely from only hours to years. Even in milder forms of the disease sudden liver failure may develop mostly as a result of portal vein occlusion. Furthermore, constantly progressing caudate lobe hypertrophy may accelerate deterioration of liver function even in patients without suprehepatic obstruction because of a decrease of functioning venous collateral pathways.

Indications

As indicated above, in almost all cases of intrahepatic obstruction of the inferior vena cava vascular stenting is one of the last palliative options. However, there is one important exception from this premise and this is in patients with Budd–Chiari syndrome associated with suprahepatic caval webs or webs below the inflow of the liver veins without evidence of engorged liver veins. Here, vascular stenting may become the treatment of first choice, since there are no further treatment options other than liver transplantation.

It is our belief that before caval stenting each patient should have meticulous investigation of the underlying disease, its metabolic disorders and the specific haemodynamic malfunction. To this end, a careful multidisciplinary approach is mandatory. As an example, metastastic liver disease compressing the inferior vena cava may be treated palliatively either by aggressive chemotherapy and even radiotherapy or caval stenting. The choice of options must be considered carefully. The indications for caval stenting are as follows:

1 Caval webs with or without Budd–Chiari syndrome.
2 Severe caudate lobe hypertrophy as a result of otherwise intractable Budd–Chiari syndrome.
3 Segmental hepatic venous obstruction in the Budd–Chiari syndrome.
4 Unresectable tumours of the caudate lobe or neighbouring segments not responding to chemotherapy.
5 Slowly or moderately progressing metastatic liver disease not responding to chemotherapy.
6 Post-transplantation strictures at venous anastomoses not amenable to surgical correction.

Contraindications

There are practically no contraindications to caval stenting with the possible exception of current sepsis. In the Budd–Chiari syndrome, venous stenting has achieved its role amongst other options such as no treatment, conservative management of symptoms attributable to portal hypertension, complex surgical shunts and liver transplantation.

Technique

The inferior vena cava may be accessed via either the transjugular or the femoral approach. The technical approach depends on the underlying situation. In cases with only caval obstruction the first step is to visualize again the baseline morphologic situation and to rule out fresh thrombus. If fresh thrombus is present this should be lysed. Then, recanalization of the occluded or stenosed section is performed. This requires similar techniques as arterial or biliary tract recanalization. Preferably, selective catheters combined with flexible wires should be used initially. If these fail stiffer wires and catheters should be used. Sometimes a combined approach from both the femoral and the right internal jugular vein is helpful. As soon as wire passage is completed and a catheter can be pushed through the obstruction measurement of the trans-stenotic pressure gradient is performed. Then, the recanalized section is predilated, preferably with the aid of low profile balloon catheters which are available in diameters up to 12 mm (e.g. Penta II, BSIC). If, initially, catheter passage is impossible because of severe friction a pull-through technique (double access) should be applied which, usually, solves this problem. As soon as predilation is completed (effacement of balloon waist) the stent of choice may be deployed. All three stent types described above have been shown to be highly effective for caval stenting. Thus, stent choice is mainly a matter of personal preference. The technique of venous stent deployment is similar to the arterial use of stents based on the same principles of complete lesion length stenting and avoidance of overstenting and understenting. In the case of the modified Gianturco stent, in particular, significant oversizing should be avoided as this may cause stent migration through the vessel wall. When using the Palmaz stent, the final diameter can be adjusted to optimize the haemodynamic conditions. A trans-stenotic gradient of less than 5 mm Hg is considered appropriate (Fig. 12.1). When infrahepatic obstructions are stented infra-atrial projection of the stent should be avoided. No complications have been reported so far with bridging of the hepatic venous inflow when this is unavoidable. The method of treating intrahepatic venous lesions associated with the Budd–Chiari syndrome is completely different: the primary approach is aimed at the intrahepatic lesion itself rather than the inferior vena cava. Depending on the lesion site transjugular or percutaneous transhepatic access is required both of which have advantages and disadvantages. Most peripheral venous lesions are best accessed from the transjugular approach which is much less dangerous than the transhepatic approach. By contrast, the transhepatic approach is probably preferable for central venous lesions. However, once transhepatic access is established successful stenting must be achieved as the risk of transperitoneal liver bleeding is excessive without substantial lowering of the intrahepatic venous pressure. For the transhepatic approach, ultrasound is quite helpful in identifying peripheral liver veins that can be used for stable access to the site of the lesion. Recanalization in this situation is much more difficult than in caval obstruction. Sometimes abundant thrombus may be present requiring direct lysis first (urokinase 125000 units/hour) and, occasionally, the lesions are so densely fibrotic that no recanalization technique other than using sharp long needles is effective. In Fig. 12.2 a case is shown in which we had to use our TIPSS needle set to recanalize a 1 cm long occlusion of the right hepatic vein after failure to pass the lesion with any kind of steerable wire or catheter.

Anticoagulation after successful stenting does not seem to be a problematic issue in caval stenting. Typically, intraprocedural heparinization with 5000 units is sufficient. When intrahepatic lesions are treated, we feel that a more stringent anticoagulation protocol is indicated: we employ full anticoagulation for three postprocedural days and three months of antiplatelet medication (100 mg aspirin/day).

Results

Initial technical success with caval stenting defined simply as complete stenting without evidence of major residual stenosis or trans-stenotic pressure gradient, is extremely high and approaches 100% in the hands of experienced interventionists. Initial clinical success is somewhat more difficult to define as the full symptom spectrum has many different elements in both benign and malignant disease. However, in virtually all cases in which technical success is achieved the acute symptoms of inferior vena cava congestion disappear within a very short period of time. Long-term success is mainly determined by the course of the causative disease. In caudate lobe tumours, in particular, stent overgrowth or ingrowth may develop at the final stages requiring a second intervention. Following successful stenting of segmental hepatic venous stenoses the impairment of liver function usually resolves completely within weeks. However, because of long-lasting caudate lobe hypertrophy a mild or moderate inferior vena cava syndrome may persist for a longer period of time. In the Budd–Chiari syndrome the one-year clinical success rate approaches 75% of all cases treated by long segment stenting of the inferior vena cava.

PORTOCAVAL STENTS

Introduction

The pathophysiology of cirrhosis and its effects on the portal venous circulation are important to considerations regarding any form of treatment of portal hypertension (Rector et al., 1988). Useful information on the effects of portosystemic shunts has emerged from surgical studies (Redeker et al., 1958; Foster et al., 1971; Reichle et al., 1979; Sarfeh et al., 1983; Rypins et al., 1984). In 1969 Rösch et al. published initial results of a mainly nonsurgically

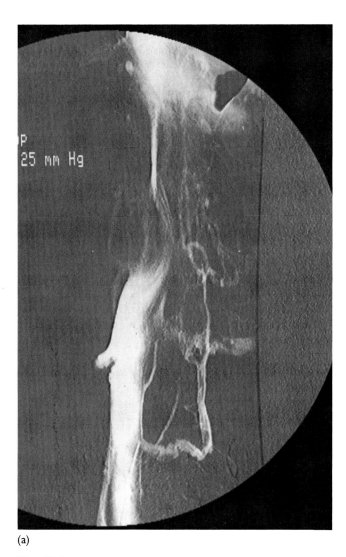

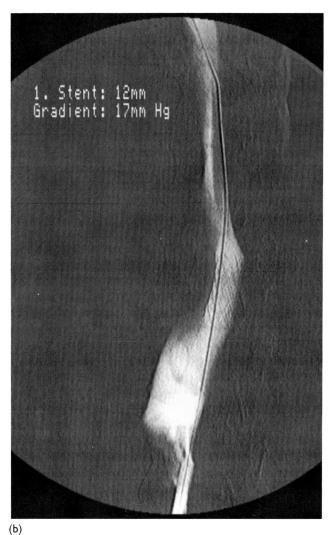

(a)   (b)

Fig. 12.1
Stenting of a patient with severe metastatic liver disease causing inferior vena cava (IVC) syndrome. (a) IVC obstruction. Collateral views are seen. (b) Following insertion of a Palmaz stent satisfactory flow has been re-established and the collaterals have disappeared.

created portosystemic connection between the inferior vena cava and the portal circulation in a canine model Rösch *et al.*, 1969. During the following decade similar models were described; however, none of them achieved acceptable clinical relevance for potential treatment of portal hypertension (Koch *et al.*, 1973; Reich *et al.*, 1977; Burgener and Gutierrez, 1979). The first realization of nonsurgical portosystemic shunting in humans was reported by Colapinto *et al.* (1982). By means of long-term balloon dilation they tried to keep patent a connection between the liver veins and the portal circulation previously established by transjugular puncture. Patency, however, was very poor. In larger series the concept proved unsatisfactory due to high rates of tract occlusion and rebleeding (Abecassis *et al.*, 1985; Gordon *et al.*, 1987). Portosystemic stents

became a practical proposition following the development of balloon-expandable metallic endoprostheses, particularly, the so-called Palmaz stent in experimental dog models in 1985 and in 1987, respectively, Palmaz demonstrated sufficient long-term patency of portosystemic shunts bridging the inferior vena cava and the portal circulation by scaffolding the intrahepatic tract with his stent (Palmaz *et al.*, 1985, 1986). Rösch *et al.* (1987), too, tried to scaffold artificial intrahepatic portosystemic tracts in a swine model with a different stent type: they used the modified Gianturco stent, a self-expandable metallic endoprosthesis. Long-term patency rates, however, were less favourable than in Palmaz's series due to shunt occlusion caused by parenchymal overgrowth. The question of whether stent design problems or a peculiar response of swine liver were

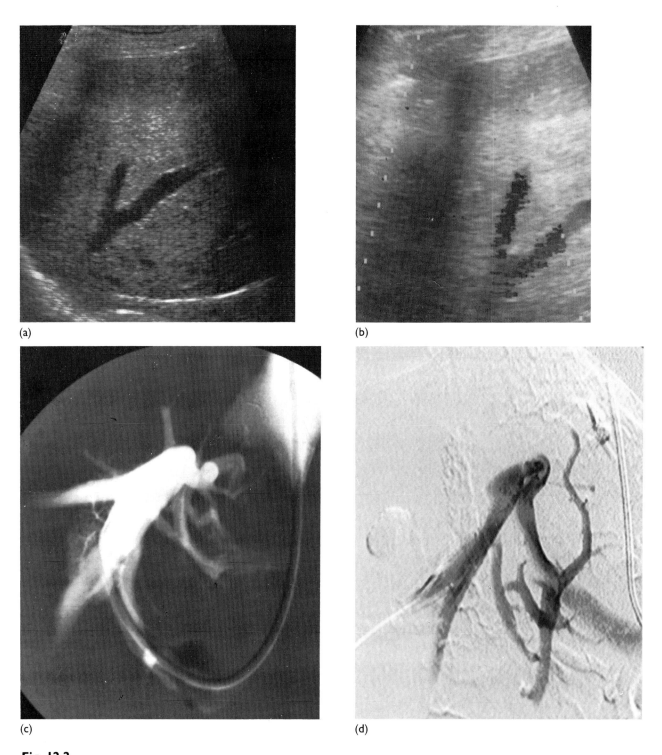

(a)

(b)

(c)

(d)

Fig. 12.2

Complex interventional approach for a patient with Budd–Chiari syndrome as a result of short segment occlusion or atresia of the right and middle hepatic vein at its caval inflow. (a) Sonographic sagittal view through the liver from a paraumbilical beam position focussing on the central liver vein radicles. Both the dorsal right and the ventral middle hepatic vein show normal diameter in the periphery and merge close to the liver dome where complete occlusion of the main stem is clearly seen. (b) Same view in the colour-coded mode. Note the well-preserved antegrade hepatic venous flow indicating well-functioning collaterals. (c) Transjugular catheterization of a hypertrophied caudate lobe vein collateralizing the occluded right and middle hepatic vein. The gradient between right hepatic vein and right atrium is 41 mm Hg and the gradient between the inferior vena cava and the right atrium is 21 mm Hg. (d) Percutaneous transhepatic approach to the right hepatic vein, again, confirming collateral flow. (e) Control after transhepatic insertion of transjugular intrahepatic portosystemic stent shunt (TIPSS) needle (Angiomed) through a 9 Fr sheath and transfemoral positioning of a Cobra-type catheter with the tip in the stump of the occluded liver vein to serve as a target for the needle. (f) Control after careful advancement of the mildly curved needle towards the tip of the catheter

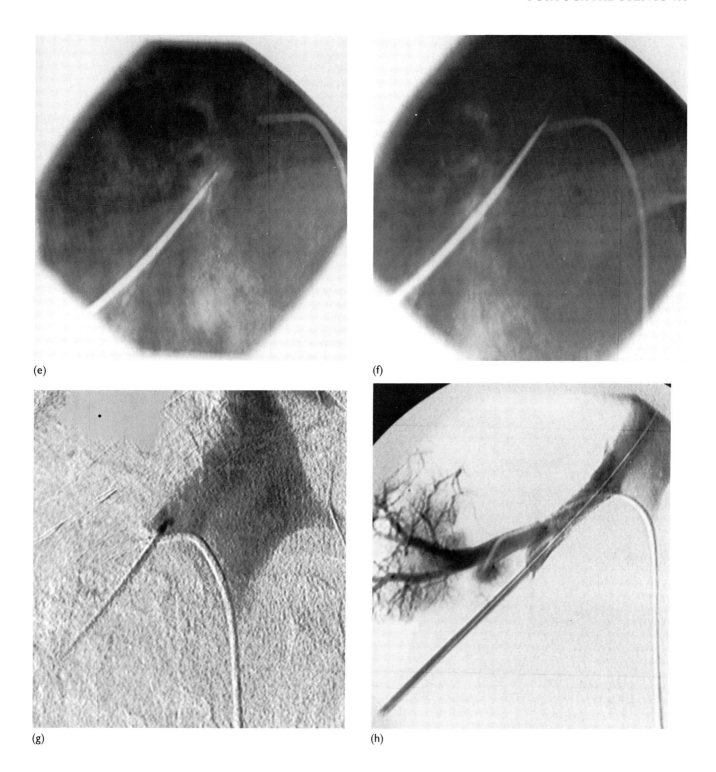

(e)

(f)

(g)

(h)

(relatively little resistance). (g) Confirmation of correct puncture of the inferior vena cava. (h) Control after stent placement of one Palmaz iliac stent with a diameter of 10 mm. Good flow across the stented hepatic vein segment. The trans-stenotic gradient is 0 mm Hg. The injection was performed through the transfemoral Cobra catheter.

the reason for the unfavourable results could not be settled (Rösch *et al.*, 1987). Palmaz achieved 50% long-term patency in a canine model without portal hypertension (Palmaz *et al*, 1985) and 100% patency when the animals had portal hypertension (Palmaz *et al.*, 1986). From this experience it was justified to seek a reproducible and reliable technique for transjugular intrahepatic portosystemic stent shunting to treat patients with severe portal hypertension. In 1987 we submitted to the Institutional Revision Board of the University Hospital of Freiburg ('Ethik-Kommittee') a pilot study application for first clinical use of TIPSS. The first successful procedure was accomplished in January 1988 (Richter *et al.*, 1989). Since then we have developed and improved the TIPSS technique based on experience gathered in more than 100 patients to date (Richter *et al.*, 1990a,b). Initially, the study protocol restricted the use of the procedure to patients in whom conservative management with repeat sclerotherapy and vasoactive drugs failed to control bleeding and who were at prohibitive risk for surgical shunts. Subsequent accumulation of technical and clinical experience encouraged a wider and more deliberate use of TIPSS to include younger patients and patients with milder stages of liver disease.

Indications

An assessment by an experienced endoscopist of the state and stages of oesophageal and gastric varices is mandatory before TIPSS is considered. In particular, a reasonable prospect of success by continuation of sclerotherapy should be ruled out. The following list of indications seems to be appropriate:

1 chronically recurring variceal bleeding despite continuing sclerotherapy.
2 Recurring variceal bleeding and severe ulcerative or erosive disease from repeat sclerotherapy.
3 Repeat bleeding episodes from major gastric varices inaccessible for sclerotherapy.
4 Recurring variceal bleeding from occluded surgical shunts.

Patients may be considered for TIPSS because the risks associated with surgery are unacceptable in certain cases. For this particular subset of patients no therapeutic choice other than TIPSS exists. For patients fit for surgery in stage A or B of the Child and Turcotte classification of liver disease (Child and Turcott, 1964) the classical therapeutic option has been the distal splenorenal shunt (DSRS, or Warren shunt). In such patients TIPSS seems to compete with surgery (Warren *et al.*, 1982). More experience

is necessary to enable stage-related and risk – stratified analysis of patency rates, risk of encephalopathy, and an assessment of the morbidity and mortality of the TIPSS method in comparison with shunt surgery.

However, at present it seems safe to state that TIPSS can be recommended for patients already enrolled on a liver transplantation waiting list who are threatened by a high bleeding risk while waiting for the life-saving organ. In these patients the performance of a TIPSS procedure leaves the main vascular structures untouched in contradistinction to shunt surgery. Previous shunt surgery is well known to increase surgical difficulties and morbidity during liver transplantation.

An issue which is still unclear is whether TIPSS should be offered to patients with intractable ascites. Our long-term results in this particular subset of patients is still too limited to allow more than preliminary conclusions to be drawn. It appears that patients in whom ascites develops rapidly and is significantly associated with a high portosystemic pressure gradient TIPSS may be very beneficial. By contrast, in patients with a long history of liver cirrhosis bordering on liver failure, and small organ size without a high portosystemic pressure gradient, TIPSS may lead to rapid liver failure from deprivation of portal blood nutrition. It is essential to measure the venous occlusion pressure in patients referred for TIPSS to treat intractable ascites. This may help to select patients who will not benefit from the procedure and should either be transferred to a liver transplantation programme or be left alone.

Contraindications

The contraindications to TIPSS are not necessarily the same as those to shunt surgery because the radiological intervention is much less invasive. However, there are four absolute contraindications:

1 Right heart failure or other cardiopulmonary factors contributing to substantial elevation of right ventricular pressure (chronic or acute left heart failure, cor pulmonale etc.).
2 Sepsis. Special attention must be paid to pulmonary infection as this may easily develop from aspiration pneumonia during bleeding episodes and sclerotherapy. Also, ascitic fluid can become infected in long-standing liver cirrhosis.
3 Significant acute liver failure not attributable to active bleeding.
4 Presence of hepatocellular carcinoma (HCC) compressing or infiltrating hepatic vessels or the

parenchyma of the liver adjacent to the proposed shunt tract.

A relative contraindication is portal vein occlusion. With wider clinical application of the TIPSS concept more experience with this problem has emerged. In a recent update the San Francisco group reported a 70% initial success rate in such patients (E. Ring, personal communication). Another relative contraindication is peripheral small HCC in patients unfit for surgical resection.

Technique

There were several complications during the first part of the learning curve with the procedure. Many of these early problems have been published in detail (Murray *et al.*, 1961; Koch *et al.*, 1973; Millikan *et al.*, 1985; Lafortune *et al*, 1987; Johansen, 1989). To avoid unnecessary overalap we shall to concentrate on our up-to-date technique which reduced the procedural time from 7 hours to a little over 2 hours in the majority of cases.

PATIENT PREPARATION

A variety of appropriate clinical and laboratory tests should be performed before TIPSS to allow the application of the Child and Turcott classification of liver cirrhosis (Child and Turcott, 1964) in the modified version according to Conn (Fischer and McClinley, 1985). Subclinical hepatic encephalopathy can be diagnosed by applying the Number Connection Test according to Conn (Delacy *et al.*, 1989). Imaging procedures are performed to rule out malignant disease, infection, portal vein occlusion and major anatomic abnormalities preventing successful performance of the procedure. Abdominal ultrasound including Doppler sonography of the portal vein, chest radiography and abdominal angiography are carried out. If HCC or portal vein occlusion are suspected additional abdominal computed tomographic (CT) studies should be performed. If the hepatic veins are well seen on CT a good appreciation of the anatomic relationship between them and the portal vein is possible.

It is important to try to improve the clinical state of each individual patient before TIPSS. The measures taken to achieve this include correction of haematocrit, protein and coagulation deficits, bowel cleaning and sterilization, drainage of ascitic fluid, correction of electrolyte imbalance and prophylactic broad-spectrum antibiotic therapy. Immediately before the procedure three to six of blood units should be cross-matched.

ANATOMIC SITUATION

We consider the creation of a wide, central and approximately straight intrahepatic shunt tract as crucial for early and long-term success. In view of this consideration, the puncture tract should bridge the proximal part of either the right or the middle hepatic vein with the upper wall of either one of the main portal vein branches. Therefore, exact knowledge of the anatomic relationship between the portal bifurcation and the hepatic vein radicles is mandatory. In most patients the portal bifurcation is located anterior to the main stems of the hepatic veins, but there are several anatomic variants. Of particular interest is the position of the liver capsule along the course of the portal vein. Inadvertent puncture of the extracapsular part of the portal vein carries a high risk of life-threatening intra-abdominal bleeding. Both CT and ultrasound scanning are extremely helpful in identifying and defining the individual anatomic situation in each patient.

TRANSJUGULAR ACCESS

For sterility reasons the procedure is performed in the angiography suite and draping after careful skin preparation. After sonographic documentation of the course of the right internal jugular vein and skin anaesthesia a bevelled 18 gauge cannula with a 5 ml syringe connected to the hub and filled with normal saline is introduced 4–5 cm cranial to the upper aspect of the clavicle at a very shallow angle to provide easy access for the large bore instruments to follow. When blood is drawn easily a 0.035 inch J guidewire is inserted and manipulated down to the inferior vena cava under fluoroscopic control to allow insertion of a long 8 F sheath (Terumo). If the wire cannot be advanced to the inferior vena cava a selective cather is used.

THREE-DIMENSIONAL ORIENTATION

Most of the difficulties associated with punctures aimed centrally and medially towards the portal bifurcation have been eliminated by combined fluoroscopic and ultrasound guidance of the puncture. However, it has to be understood that the sonographic appearance of cirrhotic liver is significantly changed and vascular structures are much less visible compared to normal liver. Hence, only state-of-the-art ultrasound technology displays the relevant structures, particularly, when colour Doppler mode imaging is used.

Sonographic guidance from an intercostal lateral view visualizes both the bifurcation of the portal vein

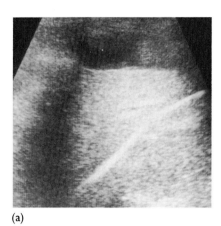

(a)

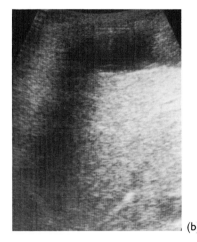

(b)

Fig. 12.3

Transjugular intrahepatic portosystemic stent shunt (TIPSS) technique: demonstration of sonographic guidewire identification. (a) Sonographic demonstration from a midaxillary lateral view of a superstiff guidewire lodged very peripherally in the segmental hepatic vein branch which divides segments t and 6 of the right liver lobe. (b) Similar sonographic beam direction: the wire is pulled back to a position from where the shunt tract should start to allow the shortest and straightest course possible.

and the hepatic veins. By directing the beam anteriorly or posteriorly the length and angulation of the shortest shunt tract between both venous systems may be easily determined. In connection with CT studies this also helps to determine the orientation of the portal bifurcation in relation to the hepatic veins.

PUNCTURE AND SHUNT TRACT CREATION

The first step after the establishment of transjugular access is catheterization of the hepatic veins. Usually, we use a 5 Fr multipurpose catheter (Terumo) and start with the right hepatic vein which in the vast majority of the cases will be the one to use. A hepatic venogram carried out with the catheter tip positioned distally in the vessel demonstrates the anatomical situation. Particular attention is drawn to the size of the vein at its inflow into the inferior vena cava. The diameter should be at least 10 mm. In some instances there is retrograde sinusoidal flow which quickly identifies the portal bifurcation. After this, a superstiff wire is inserted which helps to visualize the vein by ultrasound (Fig. 12.3). Then, an 8 Fr guiding catheter from the TIPSS set designed by us (Angiomed) is introduced over the wire. With the set comes a puncture device that has been specially designed for the TIPSS procedure. It features a blunt 50 cm long cannula which has a 15 G shaft tapered to 17.6 G at the tip. The cannula is prebent to 30° and is stiff for sufficient torque control. To function as a needle an inner mandril made from nitinol is inserted into the cannula which is extremely sharp and highly

flexible with a smooth transition from its tip to the cannula. The cannula can be inserted in the liver over the wire already in place or by using a blunt obturator which is also part of the set and also made from nitinol. In most cases we change the curve of the needle to adopt the angle at which the hepatic vein enters the inferor vena cava. If this approximates 90° heavy manual bending is required. Then, the cannula can only be inserted over a superstiff wire positioned with its stiff part well into the peripheral hepatic vein covered by the guiding catheter in order to minimize the risk of catheter perforation. With the aid of ultrasound and fluoroscopy the optimal puncture site is chosen. Usually, ultrasound helps to identify a position in the hepatic vein from which a short and straight course to the portal bifurcation can be established. Then the cannula is securely held in place and the sharp mandril inserted and locked to the hub of the cannula. The needle is rotated according to the predetermined puncture direction.

Combined use of ultrasound and fluoroscopy is used to monitor the intrahepatic advancement of needle (Fig. 12.2). As soon as the needle reaches the portal vein wall significant resistance is felt which has to be overcome with some pushing force. Correct portal access is confirmed as soon as the sharp mandril is removed and blood returns upon aspiration. Then contrast medium injection should confirm the situation. During this highly critical step the needle must be perfectly held in place. With the needle tip pointing medially a superstiff 0.035 inch wire is passed which, in the majority of the cases, travels down to the superior mesenteric vein. Over both the wire and the cannula the guiding catheter is pushed into the portal vein. This manoeuvre usually requires considerable force. Following this step stable portal access is achieved. However, in some cases the wire lodges in peripheral portal branches instead of advancing centrally, and a selective catheter has to be used to direct the wire centrally. In these cases a Terumo wire is more useful than a superstiff wire.

Once stable portal access is achieved the shunt tract is predilated to 8 mm utilizing low profile, 5 Fr angioplasty catheters. Typically upon initial inflation a balloon waist forms at the junction of the puncture tract and the portal wall. In most cases several minutes of inflation are needed for complete balloon expansion and effacement of the waist (Fig. 12.4).

STENTING AND THE HAEMODYNAMIC CONCEPT

It is generally accepted that an absolute portal pressure higher than 20 mm Hg or a portosystemic gradient grater than 15 mm Hg increase the risk of variceal bleeding. This holds true both for spontaneously occurring bleeding episodes and for recurrence of bleeding after surgical shunts. Conversely, a low portosystemic gradient accompanied by high volume shunt flow may significantly increase the risk of hepatic encephalopathy. Therefore, we prefer to lower the portosystemic gradient down to approximately 12 mm Hg. Careful measurement of the portal pressure is a crucial point in the procedure. Typically, before predilation of the shunt tract, as described above, we monitor the portosystemic gradient. Then, the tract is completely scaffolded from its entry into the portal vein up into the hepatic vein using a technique identical to arterial stenting. Stenting is performed through a special 35 cm long sheath which is also part of the TIPSS set. We use as many Palmaz stents as necessary with an overlap of several millimetres and an initial diameter of 8 mm. The flow conditions in the hepatic vein are very important (Rössle et al., 1990). The 'outflow' must be wide enough to accept the shunt flow. Therefore, it may be appropriate to stent the whole of the hepatic vein if it turns out to be too small. Upon completion of the stent shunt with an initial diameter of 8 mm the portosystemic gradient is monitored again. Unless substantial portal decompression within a range of 10–13 mm Hg is measured stepwise increase of the shunt diameter is performed by applying balloon dilation in 1 mm increments until the desired pressure level is reached. The stented segment within the hepatic vein is flared to a trumpet shape by dilating it with a 12 mm balloon in order to allow easier follow-up catheterization.

VARICEAL EMBOLIZATION

The need for embolization of varices as an adjunct to a successful TIPSS procedure in an acutely bleeding patient is an unsettled issue (Coldwell et al., 1991). In our opinion, simultaneous embolotherapy (by whatever means) in addition to the creation of a well-functioning shunt can accelerate patient recovery (Fig. 12.5). Usually, acutely bleeding patients in whom medical treatment has failed to control variceal haemorrhage present in a poor or critical clinical state resulting from substantial intestinal blood loss, coagulopathy – possibly worsened by mass transfusion – and hepatic encephalopathy. Failed vasopressive intravenous therapy and prolonged inflation times of gastro-oesophageal balloons increase the risk of an already life-threatening situation.

PATIENT MONITORING DURING TIPSS

Significant pain usually accompanies dilation of the shunt tract and stent deployment. The discomfort should be alleviated by appropriate intravenous analgesics under oxygen saturation monitoring. Inadvertent catheter or wire passage into the right ventricle frequently happens at a variety of stages during the procedure and may cause severe arhythmia. Therefore, constant ECG monitoring is required with antiarhythmic medication ready at hand.

SPECIFIC MEDICATION FOR TIPSS AND POSTPROCEDURAL CARE

In elective procedures 3–6 units of whole blood are prepared depending on coagulation status, haemoglobin level and total blood count. In cases of significant coagulation abnormalities (prolongation of the prothrombin time by more than 30% and/or the decrease of the prothrombin level to below 50% of the normal) 4–8 units of fresh frozen plasma are made available to allow instantaneous infusion if necessary to treat any bleeding complication (see also below). Broad-spectrum antibiotic therapy is started on the day of the procedure and continued for two more days. Immediately prior to stent deployment heparin is given according to the coagulation status of the patient. Patients with prothrombin levels of >60% of the control receive 5000 units, whereas those with prothrombin levels <60% and >45% receive 2500 units. Therapeutic heparinization is maintained for two more days in patients with ap-

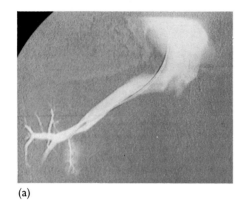

(a)

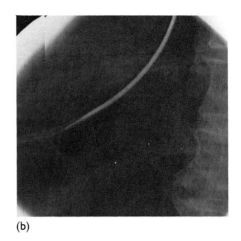

(b)

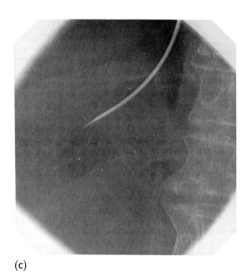

(c)

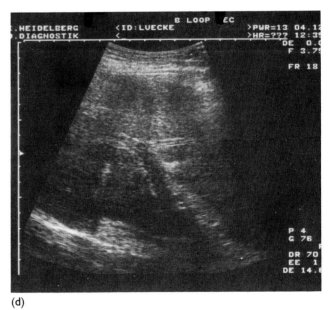

(d)

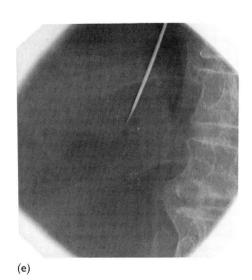

(e)

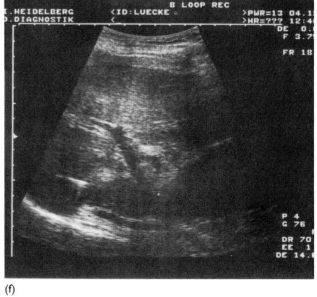

(f)

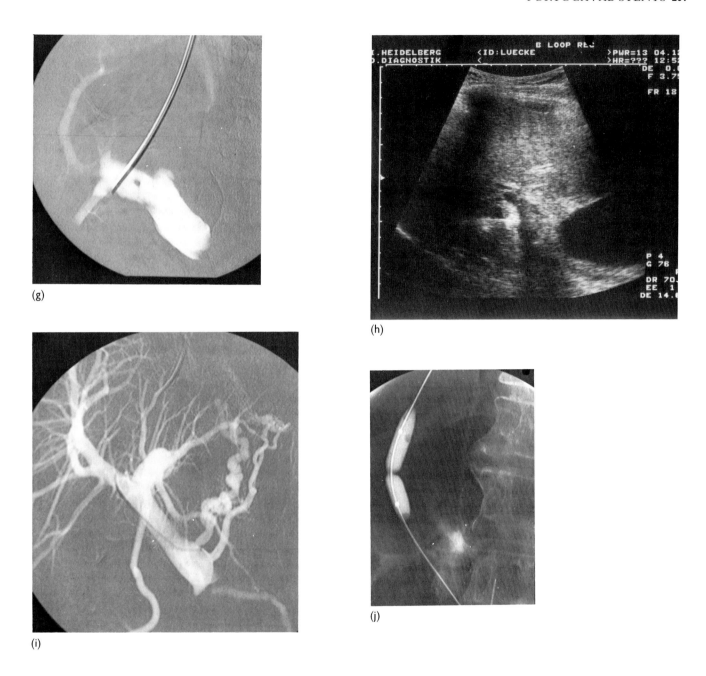

Fig. 12.4

Transjugular intrahepatic portosystemic stent shunt (TIPSS) techniques: simultaneous sonographic and fluoroscopic control of the puncture. (a) Right hepatic vein catheterization with a 5 Fr multipurpose catheter: widely patent right liver vein close to the caval inflow. (b) 8 Fr guiding catheter lodged peripherally and puncture cannula advanced with a blunt obturator (parts of the Angiomed-TIPSS set). (c) Guiding catheter pulled back and blunt obturator replaced by sharp puncture mandril locked to the hub of the cannula. (d) Sonographic demonstration of the needle tip corresponding to the fluoroscopically shown position. (e) Needle rotated medially towards the portal bifurcation. (f) Same situation demonstrated sonographically. (g) Successful puncture of the right portal branch. (h) Same situation demonstrated sonographically. (i) Direct portography after insertion of a 5 Fr catheter into the portal vein. The pressure gradient between the portal vein and the right atrium is 27 mm Hg. Significant filling of coronary vein branches. (j) Predilation of the shunt tract using a 5 Fr 8 mm angioplasty balloon (Penta II, BSIC), note the high grade waist of the fully inflated balloon (approximately with 8 atm).

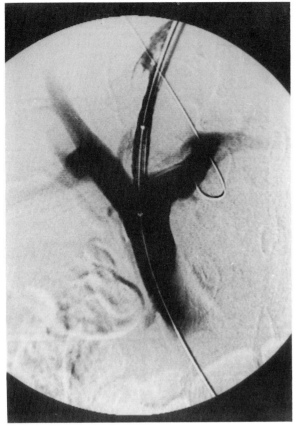

(a)

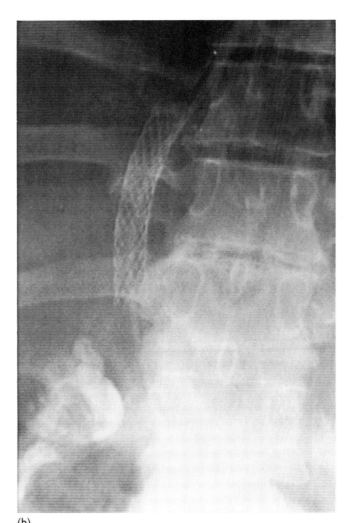

(b)

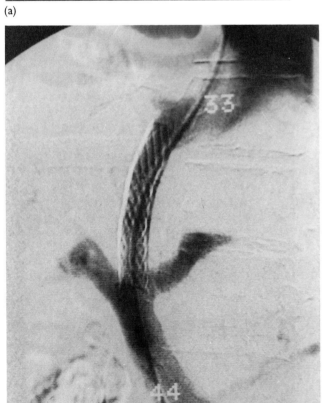

(c)

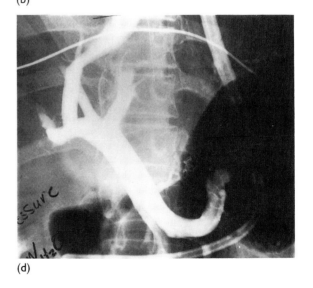

(d)

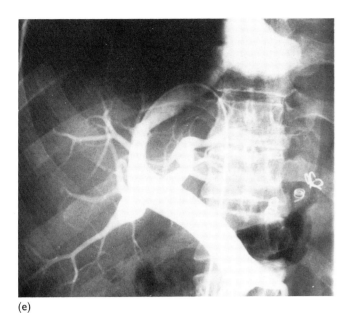

(e)

Fig. 12.5
Demonstration of the Palmaz stent deployment process and variceal embolization. (a) Dilation of tract before stenting. (b) The Palmaz stent is shown following deployment. (c) Excellent flow is seen through the stent. The gradient across the shunt was 11 mm Hz. (d) Bleeding from the oesophageal vain. This was embolized with steel coils. (e) Direct portogram (via a femoral approach) one week later shows that the varix has been occluded.

proximately normal coagulation. In these antiplatelet medication is established for three months.

In emergency shunting the full range of measures employed in the control variceal haemorrhage (occlusion tubes, β-blockers and vasoconstrictor infusions) is employed prior to the procedure. General anaesthesia is required in patients in danger of aspiration or those with severe encephalopathy who are unable to cooperate during the procedure. In acutely bleeding patients broad-spectrum antibiotic therapy is initiated at least one day before the procedure as well as medication and measures for mechanical and biologic clearing of the bowel from blood and bacteria.

After successful completion of TIPSS the patients are kept in the intensive care unit until they are clinically stable without signs of gastrointestinal haemorrhage, pulmonary infection or renal or hepatic malfunction.

In the first few days following the procedure the oesophageal varices are assessed endoscopically to compare the appearance with the situation before TIPSS. If no reduction in size is observed and there are no signs of hepatic encephalopathy are present the patient is scheduled for shunt redilation which is easily performed by simple re-expansion of the stent shunt with bigger balloons.

Normal nutrition is allowed for patients with near normal liver function. In patients with abnormal liver function a low protein diet is instituted.

Direct portography is performed as part of our routine follow-up examination programme 3, 6 and 12 months after TIPSS to examine the healing pattern of the stent shunt and observe possible onset of intimal hyperplasia (see below).

Results

CHANGES IN STUDY POPULATION AND SUCCESS RATES

To date, more than five years have elapsed since our first TIPSS (Richter *et al.*, 1989). Initially, only very high risk patients were selected for the procedure. Nevertheless, with promising results emerging during our first year of clinical application (Richter *et al.*, 1990a,b) the indications for TIPPS were extended to include patients with failed surgical shunts, those refractory to medical treatment for severe ascites, and patients on a waiting list for liver transplantation. With broadening of the inclusion criteria during our study difficulties arise in the assessment of clinical benefit and long-term success of the procedure. In our first 24 patients, the technical success rate was 75% (Richter *et al.*, 1991). Failures resulted from inability to puncture the portal vein because of equipment problems as detailed above. Following improvements in technology and technique the overall success rate is now 92% and the technical success rate in 1992 was 97%.

THIRTY-DAY MORTALITY AND COMPLICATIONS

The 30-day mortality rate in the first 13 (out of 18) successful cases was 15% (Richter et al., 1990a,b, 1991). This has completely changed since the advent of improved equipment and better puncture techniques (see above). Among more than 100 procedures completed by simultaneous fluoroscopic and sonographic guidance four early deaths occurred, only one of which was directly related to the procedure. This death resulted from inadvertent puncture of extracapsular portal vein which led to exsanguination despite successful surgical emergency shunt. Two deaths resulted from septic complications. The fourth patient died from malignant infiltration of the portal and mesenteric vein because of previously undiagnosed HCC.

Other complications were much less severe and included four haematomas at the entry site to the internal jugular vein. Seven patients had transient elevation of bilirubin and transaminases without unfavourable sequelae. Two patients showed signs of mild haemolysis which could not be adequately explained. In six patients balloon rupture upon stent placement required sophisticated stent correction methods all of which were successful.

THIRTY-DAY CLINICAL SUCCESS

The total early clinical success rate is 93.3%; rebleeding occured in 6.7% (8/112) during the 30-day period. Five patients rebled from pre-existing severe ulcerative and erosive oesophageal and gastric mucosal disease probably caused by extensive sclerotherapy trials accompanied by prolonged inflation times of occlusion balloons. Our very first patient in the series developed widespread mucosal bleeding two days after TIPSS most likely due to disseminated intravascular coagulopathy caused by accelerated absorption of ascites. This was easily controlled by blood transfusions and fresh frozen plasma (Richter et al., 1989). In two other patients bleeding episodes continued for approximately two weeks.

THIRTY-DAY ENCEPHALOPATHY

In addition to mortality and shunt occlusion hepatic encephalopathy is one of the main complications of shunt surgery (Warren et al., 1982; Galambos, 1985; Millikan et al., 1985; Ohnishi et al, 1985; Sarfeh et al., 1986; Lafortune et al., 1987; Spina et al., 1988;

Johansen, 1989; Pagliaro et al., 1989) and it is also a crucial issue in TIPSS. In nonselective shunts the postoperative rate of hepatic encephalopathy may rise to 50% (Palmaz et al., 1986; Sarfeh et al., 1986; Richter et al., 1991). Even in selective shunts an incidence of hepatic encephalopathy as high as 20% has been reported (Sarfeh et al., 1986; Richter et al., 1991). In our series, six patients developed hepatic encephalopathy during the first 30 days which was controlled by appropriate medical treatment in each case. More importantly, in none of our 39 stage C patients who almost invariably presented with hepatic encephalopathy before TIPSS did worsening of symptoms occur. Those patients who had hepatic encephalopathy attributable to severe acute bleeding and significant intestinal protein uptake improved as soon as the shunt was established. These findings are reflected in variations in ammonia levels. In most of the patients with normal values before TIPSS they did not exceed the critical threshold during follow-up. In patients with increased values due to active bleeding ammonia levels decreased after TIPSS in most of the cases. The series is still too small for firm conclusions to be drawn regarding the precise risk of hepatic encephalopathy in TIPSS. The relatively low incidence of de novo hepatic encephalopathy should be considered as a positive trend favouring the concept of partial diversion of portal flow volume with small calibre interposition shunting.

LATE RESULTS

The actuarial one-year survival rate is 70% (44/63), and the three-year survival rate 50% (12/24). Presently, 89 patients are living with an average survival time of 13 months. In addition to the previously mentioned seven early deaths another 16 patients died with an average survival time of 10 months. Death was unrelated to the procedure in 13 cases while in two patients late shunt occlusion occured 9 and 18 months, respectively, after TIPSS resulting in lethal variceal haemorrhage.

Two patients were referred for liver transplantation because of progressive deterioration of liver function during a period of five months.

De novo encephalopathy was seen in three patients during late follow-up mostly attributable to failure to respond to protein restriction. In all patients adequate hydration and re-establishment of correct dietary schedules restored normal brain function.

Intimal hyperplasia during the first six months after TIPSS is an important feature in patients with good liver function and normal or close-to-normal coagulation. In almost all patients in a Child's A stage of liver disease significant intimal hyperplasia was seen both within the stented shunt segments and in the free area of the hepatic vein. It was never seen in the portal vein. When haemodynamically necessary, as determined by portosystemic pressure gradient monitoring, correction by either redilation or additional stenting of hepatic vein segments may be easily achieved. By contrast, in patients with reduced liver function such intimal hyperplasia is rare. This underlines the thrombogenic property of metallic stents particularly in areas without functioning endothelium and well-functioning blood coagulation.

Thrombogenicity and haemodynamically significant intimal hyperplasia are some of the unsolved problems of the TIPSS procedure which warrant further research.

THE RATIONALE OF STENT CHOICE FOR TIPSS

The flexible Wallstent can stent more peripheral and curved shunt tracts allowing a 'take-what-you-can-get' policy after less than ideal punctures. With the use of the more rigid Palmaz stent a more central and straight intrahepatic shunt course is mandatory requiring a more accurate puncture. Nevertheless, we believe that a central and straight shunt tract facilitates an undisturbed flow pattern leading to as little as possible thrombus formation and intimal hyperplasia. This is important during shunt maturation and stented tract healing. Most of the difficulties associated with punctures aimed centrally and medially towards the portal bifurcation have been almost completely eliminated by the use of direct ultrasound guidance in addition to fluoroscopy. The radio-opacity and expansion mechanism of the Palmaz stent allows precise positioning within the target area (Figs 12.3 and 12.4). Excessive protrusion of the stent inside portal venous or hepatic venous structures may be avoided. By Contrast, the Wallstent substantially extends into both the portal and the hepatic veins because of its unpredictable deployment and site of foreshortening. At present it is not known whether such protrusion of metallic stents inside portal or hepatic veins may have adverse effects.

The use of the Palmaz stent allows adaptation of the shunt dimensions to the particular haemodynamic situation in each patient because a range of diameters between 7 and 16 mm is obtained by choosing the appropriate size. Notwithstanding the above-mentioned theoretical considerations the choice of stent will be determined by the personal preference of the operator in the short term and the results of larger scale clinical trials in the long term.

REFERENCES

ABECASSIS M, GORDON JD, COLAPINTO RF et al. (1985). The transjugular intrahepatic portosystemic shunt (TIPS): an alternative for the management of life-threatening variceal hemorrhage. *Hepatology* **5**: 1032A.

BURGENER FA, GUTIERREZ OH (1979). Non-surgical production of intrahepatic portosystemic venous shunts in portal hypertension with the double lumen balloon catheter. *Fortschr Roentgenstr* **130**: 686–8.

CHILD CG, TURCOTT JG (1964). Surgery and portal hypertension. In: *The liver and portal hypertension*. Edited by Child CG. Saunders, Philadelphia, 1964.

COLAPINTO RF, STRONELL RD, BIRCH SJ et al. (1982). Creation of an intrahepatic portosystemic shunt with a Grüntzig balloon catheter. *Canadian Medical Journal* **126**: 267–326.

COLDWELL DM, MOORE ADA, BEN-MENACHEM Y, JOHANSEN KH (1991). Bleeding gastroesophageal varices: gastric vein embolization after partial decompression. *Radiology* **178**: 249–51.

DELACY AM, NEVASA M, GARCIA-PAGAN JC et al. (1989). Reversal of portal flow after distal splenorenal shunt (DSRS). Relationship to hepatic encephalopathy and impaired liver function. *Journal of Hepatology* **9** (Suppl): S142.

FISCHER JE, CCCINLEY J (1985). Comparative randomized study: Proximal versus distal splenorenal shunt. *Policlinico Sez Chir* **92**: 592–6.

FOSTER JH, ELLISON LH, DONOVAN TH, ANDERSON A (1971). Quantity and quality of survival after portosystemic shunts. *American Journal of Surgery* **12**: 490–501.

GALAMBOS JT (1985). Portal hypertension. *Seminars in Liver Disease* **5**: 277–90.

GORDON JD, COLAPINTO RF, ABECASSIS M et al. (1987). Transjugular intrahepatic portosystemic shunt: A nonoperative approach to life-threatening variceal bleeding. *Canadian Journal of Surgery* **30**: 45–9.

JOHANSEN K (1989). Partial portal decompression for variceal hemorrhage. *American Journal of Surgery* **157**: 479–82.

KOCH G, RIGLER B, TENTZERIS M et al. (1973). Der intrahepatische porto-cavale Shunt. *Langenbecks Arch Chir* **333**: 237–44.

LAFORTUNE M, PATRIQUIN H, POMIER G et al. (1987). Hemodynamic changes in portal circulation after portosystemic shunts: Use of duplex sonography in 43 patients. *American Journal of Radiology* **149**: 701–6.

MILLIKAN WJ, WARREN WD, HENDERSON JM et al. (1985). The Emory prospective randomized trial: selective versus non-selective shunt to control variceal bleeding. *Annals of Surgery* **201**: 712–22.

MURRAY JF, MULDER DG, NEBEL L (1961). The effect of

retrograde portal venous flow following side-to-side portocaval anastomosis. *Journal of Clinical Investigation* **40**: 1413–20.

OHNISHI K, SAITO M, SATO S et al. (1985). Direction of splenic venous flow assessed by pulsed Doppler flowmetry in patients with large splenorenal shunts. Relation to spontaneous hepatic encephalopathy *Gastroenterology* **89**: 180–9.

PAGLIARO L, BURROUGHS AK, SORENSEN TIA, LEBREC D, MORABITO A, D'AMICO G, TINÉ F (1989). Therapeutic controversies and randomised controlled trials (RCTs): prevention of bleeding and rebleeding in cirrhosis. *Gastroenterology International* **2**: 71–84.

PALMAZ JC, GARCIA F, SIBBIT SR, TIO FO, KOPP DT, SCHWESINGER W, LANCASTER JL, CHANG P (1986). Expandable intrahepatic portacaval shunt stents in dogs with chronic portal hypertension. *American Journal of Radiology* **147**: 1251–4.

PALMAZ JC, SIBBITT RR, REUTER SR, GARCIA F, TIO FO (1985). Expandable intrahepatic portacaval shunt stents: early experience in the dog. *American Journal of Radiology* **145**: 821–5.

RECTOR WG, HOEFS JC, HOSSACK KF, EVERSON GT (1988). Hepatofugal portal flow in cirrhosis: Observation of hepatic hemodynamics and the nature of the arterioportal communications. *Hepatology* **8**: 16–20.

REDEKER AG, GELLER HM, REYNOLDS TB (1958). Hepatic wedge pressure, blood flow, vascular resistance and oxygen consumption in cirrhosis before and after end-to-side portocaval shunt. *Journal of Clinical Investigation. Journal of Surgical Research* **37**: 606–18.

REICH M, OLUMIDE F, JORGENSEN E, EISEMAN B (1977). Experimental cryoprobe production of intrahepatic portacaval shunt **23**: 14–18.

REICHLE FA, FAHMY WF, GOLSORKHI M (1979). Prospective comparative clinical trial with distal splenorenal and mesocaval shunts. *American Journal of Surgery* **137**: 13–21.

RICHTER GM, PALMAZ JC, NÖLDGE G, RÖSSLE M, SIEGERSTETTER V, FRANKE M, WENZ W (1989). Der transjuguläre intrahepatische portosystemische Stent-Shunt (TIPSS). *Radiologe* **29**: 406–11.

RICHTER GM, NOELDGE G, PALMAZ JC, ROESSLE M (1990a).

The transjugular intrahepatic portosystemic stent-shunt (TIPSS): results of a pilot study. *Cardiovascular Interventional Radiology* **13**: 200–7.

RICHTER GM, NOELDGE G, PALMAZ JC, ROESSLE M, SIEGERSTETTER V, FRANKE M, GEROK W, WENZ W, FARTHMANN E (1990b). Transjugular intrahepatic portacaval stent shunt: preliminary clinical results. *Radiology* **174**: 1027–30.

RICHTER GM, NOELDGE G, PALMAZ JC, ROESLLE M (1991). Evolution and clinical introduction of TIPSS, the transjugular intrahepatic portosystemic stent-shunt. *Seminars in Interventional Radiology* **8**: 331–40.

RÖSCH J, HANAFEE WN, SNOW H (1969). Transjugular portal venography and radiologic portocaval shunt: an experimental study. *Radiology* **92**: 1112–14.

RÖSCH J, UCHIDA BT, PUTNAM JS et al. (1987). Experimental intrahepatic portocaval anastomosis: use of expandable Gianturco stents. *Radiology* **162**: 481–5.

RÖSSLE M, HAAG K, NOELDGE G, RICHTER G, WENZ W, FARTHMANN E, GEROK W (1990). Hämodynamische Konsequenzen der portalen Decompression: Welches ist der optimale Shunt? *Z Gastroenterol* **28**: 630–4.

RYPINS EB, MASON, GR, CONROY RM, SARFEH IJ (1984). Predictability and maintenance of portal flow patterns after small-diameter portocaval H-grafts in man. *Annals of Surgery* **200**: 706–10.

SARFEH IJ, RYPINS EB, CONROY RM, MASON GR (1983). Portocaval H-graft: relationships of shunt diameter, portal flow patterns and encephalopathy. *Annals of Surgery* **197**: 422–6.

SARFEH IJ, RYPINS EB, RAISZADEH M, MILNE N, CONROY RM, LYONS KP (1986). Serial measurement of portal hemodynamics after partial portal decompression. *Surgery* **100**: 52–8.

SPINA GP, GALEOTTI F, OPOCHER E, SANTAMBROGIO R, CUCCHIARO G, LOPEZ C, PEZZUOLI G (1988). Selective distal splenorenal shunt versus side-to-side portocaval. Clinical results of a prospective controlled study. *American Journal of Surgery* **155**: 564–71.

WARREN WD, MILLIKAN WJ JR, HENDERSON JM et al. (1982). Ten years portal hypertensive surgery at Emory: results and new perspectives. *Annals of Surgery* **195**: 530–42.

Index

Abscess(es)
  fistula-associated, drainage 161–5
  intra-abdominal, drainage *see* Drainage
  intrahepatic
    percutaneous biliary drainage complicated by 29
    percutaneous biliary drainage in presence of 20
    stenting complicated by, endoscopic/percutaneous
      biliary 87
  subphrenic, percutaneous biliary drainage
    complicated by 29
Achalasia, dilation 133, 135
Ampullary carcinoma, management 91
Anaesthesia, strictures and 54
Anastomotic strictures
  biliary–enteric 53, 57, 59, 60
  gastrointestinal 147–58 *passim*
  oesophageal 132, 133, 137, 138
Angiography, embolization employing 176, 185, 187
Angioplasty balloon catheters *see* Balloon catheters
Antibiotics
  in gastrointestinal fistula management 160
  prophylactic
    endoprosthesis/stent insertion using 36, 84
    endoprosthesis/stent patency testing using 50
    percutaneous biliary drainage using 20
Anticoagulants, transjugular intrahepatic
    portosystemic stent shunts and use of 199
Appendectomy, enterocutaneous fistula following 162
Arteriovenous malformations, hepatic 177, 184, 185
Ascites
  percutaneous biliary drainage in presence of 20
  transjugular intrahepatic portosystemic stent shunts
    in presence of 196
Aspiration biopsy, percutaneous needle
  diagnostic 110
  in malignant obstruction 19
  needles used 101
Atrophy, lobar, percutaneous biliary drainage in
    presence of 20

Bacteraemia, gastrointestinal stricture dilation and
    158
Balloon
  in gastrointestinal stricture dilation
    length 137
    number/size 149
  occlusion, biliary stone manipulation with 8
  in oesophageal stricture dilation
    choice 137
    length 135
Balloon catheters, angioplasty
  endoprosthesis dislocation and use in biliary tract of
    42

portosystemic shunt tract predilated with,
    transjugular intrahepatic 199
stricture dilation with 60, 133–9, 141–58
  biliary 60
  gastrointestinal 141–58
  oesophageal 133–9
  selection 136–7
  types/specifications 60, 135
Balloon-expandable metallic stents
  biliary 34
  vascular, in liver disease 190
Baskets, stone removal
  with biliary tract calculi 4, 5, 6–7, 8, 64
  with gallbladder stones 72
Bile
  endoprosthesis encrustation with 33
  leakage around drainage catheter 29
Bile ducts *see* Ducts
Bilirubin, serum, in stent functional assessment 85
Biopsy, percutaneous needle 100–108, 118–26 *see also*
    Needles
  intra-abdominal 100–108
    route, choice 101
  liver 118–26
    in coagulative disorders 118–26
    plugged 120–124
  malignant biliary obstruction 19
Bioptome in transcaval tumour biopsy 124–5
Biopty guns 103
  plugged liver biopsies and 124
Bleeding *see* Haemorrhage
Blood vessel stents *see* Stents, vascular
Bougienage 133
Bowel/intestine *see also specific bowel regions*
  catheter/wire advancement 144–5, 146
  embolotherapy 177, 178, 183, 187
  fistulae 160, 161, 163
    effluent from, control/diversion 161
  strictures 149, 152–3, 153–4, 157–8, 158
Budd–Chiari syndrome, stenting 191, 192, 194–5
Buscopan use in stenting 84

Calculi *see* Stones
Cancer *see also specific tumours types*
  bile duct
    atrophy associated with 20
    fine needle aspiration 19
    obstruction/strictures caused by *see* Obstruction;
      Strictures, biliary
    percutaneous transhepatic cholangiography in
      15
    radiotherapy, internal 30
    stenting 41, 82–93

hilar, extending into liver, endoprosthetic occlusion
    in advanced 48
liver
  embolotherapy 176–7, 179–82, 184–6
  inferior vena caval invasion by, biopsy 124–6
  percutaneous biliary drainage in presence of 20
  portosystemic stent shunt in presence of,
    transjugular intrahepatic 196, 197
  secondary *see* Metastases
oesophageal, strictures associated with 133, 134
pancreatic, embolotherapy 177
Cannulas/cannulation *see* Catheterization; Catheters
Carcinoma *see also* Cancer
  ampullary 91
  bile duct, intrahepatic *see* Cholangiocarcinoma
  liver
    embolotherapy 176, 181, 184–6
    portosystemic stent shunt in presence of,
      transjugular intrahepatic 196, 197
  strictures caused by 82–93
    biliary 82–93
    oesophageal 134
Cardiac disease, transjugular intrahepatic
    portosystemic stent shunt contraindicated in
    196
Carey–Coons endoprosthesis 25
  occlusion, management 43, 44–5
Catheter(s)/cannulas
  balloon *see* Balloon catheters
  in biliary tract 18–31, 42, 60
    drainage *see* Drainage catheters
    manipulation 24
    percutaneous 18–31
      indications 18
    steerable, stone extraction procedures employing
      4, 5
    transhepatic 18–31
  in embolization therapy 178–84 *passim*
  intra-abdominal drainage, percutaneous 110–111,
    111
    removal 110
    selection 110–111
Catheterization/cannulation
  for gastrointestinal fistulae 161, 162
  for gastrointestinal strictures 142–9
  for transjugular intrahepatic portosystemic stent
    shunt 198–9
Caustic substance, oesophageal strictures caused by
    ingestion of 129, 130–131, 137–8
Celestine device 133
Chemical dissolution
  biliary stones 11, 64
  gallbladder stones 77–8

Chemotherapy (chemoperfusion), liver cancer 179, 180, 181, 184
Cholangiocarcinoma
  atrophy associated with 20
  fine needle aspiration 19
  percutaneous transhepatic cholangiography in 15
  radiotherapy, internal 30
  stenting 41
Cholangiography
  endoscopic retrograde *see* Endoscopic retrograde cholangiography
  percutaneous transhepatic *see* Percutaneous transhepatic cholangiography
  in sclerosing cholangitis 94
  stone removal procedures employing 4, 5–6
  of strictures 53–4, 85, 86, 87
Cholangitis 93–4
  percutaneous biliary drainage complicated by 29
  sclerosing 93–4
    stricture management in 58–9, 93–4
  stenting complicated by 88
  suppurative, percutaneous biliary drainage in 24
Cholecystectomy, surgical 68
Cholecystitis
  acute percutaneous cholecystostomy in 68–71
  poststenting 87, 88
Cholecystolithotomy, percutaneous 71–7
Cholecystostomy, percutaneous 68–71
  in cholecystitis (acute) 68–71
  in gallstone management 72
Choledochoscopy
  biliary calculus lithotripsy employing 9–11
  gallbladder stone extraction employing 72
Cholesterol-containing stones, dissolution 11, 64, 77–8
Cis-platinum, liver cancer 179, 180, 181, 184
Clotting profile *see also* Coagulation abnormalities
  percutaneous biliary drainage and 19
  percutaneous transhepatic cholangiography and 14
Coagulation abnormalities *see also* Clotting profile; Haemorrhage
  cholangiography regarding, percutaneous transhepatic 14
  liver biopsy with, percutaneous 118–26
  transjugular intrahepatic portosystemic stent shunts regarding 199
Co-axial method of biopsy needle placement 101–2
Cobra catheters in pancreatic embolization 183
Coils, embolization, in liver 121–4, 179, 183
Colon *see also* Bowel; Large bowel
  embolotherapy 187
  fistulae 165
  hepatic flexure of, percutaneous biliary drainage complicated by puncture of 29
  strictures 149, 151–2, 157–8, 158
Common bile duct obstruction, percutaneous biliary drainage in 20
Computed tomography
  biopsy guidance via, percutaneous needle 103, 105–7
  cholangiography procedure employing, percutaneous transhepatic 14, 15, 16
  drainage procedure employing, percutaneous 109–10, 112, 113
    biliary 19–20
  fistulas assessed via 159
  gall stone solubility assessed via 77
  liver cancer embolotherapy 179
  strictures assessed via 54, 63
Contrast media
  biliary drainage catheter positioning and leakage of 27
  in fistulography 159–60
  in percutaneous transhepatic cholangiography 15–16
Cope biliary drainage catheters 26–7
Cope nephrostomy 27
Cope system for introducing catheters 23–4
Core-biopsy needles 101
Crohn's disease, fistulae in 165, 169
Cutaneous problems/issues *see* Skin
Cystic duct stones, impacted, removal 73, 76

Deaths, transjugular intrahepatic portosystemic stent shunt-related 203–4
Decompression, biliary 19
  imaging in determining approach to 19–20
Dilation
  balloon *see* Balloon catheters
  of bile duct
    in gallstone management 72
    stricture-associated 54
  of biliary stricture 60–61
    complications 65
    stenting after 60–61, 85
    technique of 60
  of gastrointestinal stricture 141–58
  of oesophageal stricture 133, 133–9
Dislodgement/dislocation/migration
  catheter 29
  endoprosthesis 33–4, 42–3
    prevention 33–4
Dissolution
  biliary stones 11, 64
  gallbladder stones 77–8
Drainage
  biliary, endoscopic 91
  biliary, percutaneous 18–31, 91
    endoprosthesis insertion following 28, 36
    endoprosthesis vs 28–9, 32
  catheters used *see* Drainage catheters
  of gastrointestinal fistula-associated abscesses, percutaneous 161–5
  of intra-abdominal fluid/abscess, percutaneous 108–16
  technique 110
Drainage catheters, biliary 26–9
  choice and positioning 26–7
  fixing 28
  insertion 22–5
    problems on 27–8
  number 27–8
  problems 29
Duct(s), bile *see also specific ducts*
  dilation *see* Dilation
  obstruction/strictures *see* Obstruction; Strictures
  perfusion, stone/stone fragments cleared via 64
  puncture for percutaneous biliary drainage, sites 22
  stones in *see* Stones
Duodenoscope, stenting 85
Duodenum
  embolotherapy 177–8, 178, 183
    indications/contraindications 177–8
    materials/techniques 183
    patient preparation 178
    results 186, 187
  fistula 165, 167
  perforation
    by endoprostheses (stents) 42
    percutaneous biliary drainage complicated by 29

Eder–Puestow device 133
Electrohydraulic lithotripsy of gallstones 73–6
Embolization (therapeutic) 175–88
  gastrointestinal 177–8, 178, 183, 184, 186–7
    complications/mistakes and precautions 184
    indications/contraindications 177–8
    materials/techniques 183
    patient preparation 178
    results 186–7
  gastro-oesophageal varices 199
  liver 176–7, 178, 178–82, 184–6
    of biopsy tracks 120–124, 124
    complications/mistakes and precautions 183–4
    indications/contraindications 176–7
    materials/techniques 179–82
    patient preparation 178
    results 184–6
    syndrome following 184
  pancreatic 177, 178, 182–3, 184, 186
    complications/mistakes and precautions 184
    indications/contraindications 177
    materials/techniques 182–3
    patient preparation 178
    results 186

Encephalopathy, hepatic, transjugular intrahepatic portosystemic shunts and 199, 204
Endoscopic biliary drainage 91
Endoscopic biliary stents 19, 33, 81–96
  percutaneous and, combined 92–3
  percutaneous vs 82
Endoscopic oesophageal stricture dilation 133
Endoscopic retrograde cholangiography 14
  strictures 53
Endoscopic sphincterotomy, biliary stone removal via 11
Endoscopic variceal assessment 203
Enterocutaneous fistulae 160, 161
Enteroenteric strictures 149
Epigastrium, left lobe (percutaneous biliary) drainage approach via 25, 26
Ethibloc in embolization
  of gastrointestinal tract 183
  of liver 179, 180, 181, 182, 184
  of pancreas 182–3
Evacuation of cavity (abscess) 111
Expandable stents *see also specific types*
  biliary 34, 89
    replacement 89
  vascular, in liver disease 190
Extracorporeal lithotripsy *see* Lithotripsy

Fever, poststenting 87–8
Fine needle aspiration *see* Aspiration biopsy
Fistulae, gastrointestinal 158–69
  aetiology 165–8
  location 165
  management 160–169
    conservative 160, 165
    percutaneous techniques 160–165, 165–9
  output 165
Fistulography, contrast medium 159–60
Fluoroscopy
  biopsy guidance via, percutaneous needle 103, 107–8
  drainage procedures employing guidance by 109–10
  gallstone removal by forceps employing 73, 76
  portosystemic stent shunt guidance by, transjugular intrahepatic 198, 200
  stricture access employing
    biliary 54
    gastrointestinal 141–58
    oesophageal 135, 138
Forceps, stone extraction employing
  biliary 9
  gallbladder 73

Gallstones (biliary tract/gallbladder stones) *see* Stones
Gallbladder 67–80
  percutaneous procedures 67–80
  puncture, percutaneous biliary drainage complicated by 29
Gastric embolization 177, 183
Gastric fistula 165, 166
Gastric intubation, gastrointestinal strictures and 142
Gastric strictures 154 *see also* Pylorus
Gastric varices *see* Varices
Gastric wall, pancreatic pseudocyst drainage route via 111–12, 113
Gastroduodenal artery (and branches), embolotherapy and the 178, 183, 187
Gastroenteric strictures 147–9, 154–7, 157
Gastroenterostomy, surgical, strictures following 147–9, 157
Gastrografin in post-stricture dilation oesophagogram 137
Gastrojejunal bypass, gastric outlet obstruction following 148–9, 150
Gastrostomy, pancreatic pseudocyst drainage procedure similar to 111–12, 113
Gastrostomy tube
  gastrointestinal stricture access via 142–3, 144
  oesophageal stricture patient presenting with 137
Gelfoam 124
Gianturco endoprosthesis
  Rösch modification *see* Rösch-modified Gianturco stent
  strictures treated with 35, 61
  unmodified 35

Glide wires
  in occluded metallic stent management 43–8, 48
  in strictures
    of oesophagus 136
    of pylorus 147
Gluteal muscle, drainage of pelvic fluid collections via 116
Guidewires
  in percutaneous biliary drainage
    introduction 22–5
    kinking 27
  in stricture dilation
    in gastrointestinal tract 142, 147
    in oesophagus 136

Haemodynamics, transjugular intrahepatic portosystemic stent shunt 199, 204, 205
Haemorrhage/bleeding *see also* Coagulation abnormalities
  embolotherapy 120–124, 124, 177
  percutaneous biliary drainage complicated by 29
  stenting complicated by 88
  transjugular intrahepatic portosystemic stent shunts complicated by 196, 199
  variceal 196, 199
Haemostatic protein/polymer sheath 124
Heart disease, transjugular intrahepatic portosystemic stent shunt contraindicated in 196
Hepatic arteries, embolotherapy and the 179, 184
Hepatic flexure of colon, percutaneous biliary drainage complicated by puncture of 29
Hepatocellular carcinoma (HCC)
  embolotherapy 176, 181, 184–6
  portosystemic stent shunt in, transjugular intrahepatic 197
    contraindication 196
Hilar duct
  malignancy, extending into liver, endoprosthetic occlusion in advanced 48
  obstruction/stricture 19, 91–2, 92–3
    dilation 85
    percutaneous biliary drainage 21
    percutaneous transhepatic cholangiography 17
    stenting 85, 91–2, 92–3
Holes, side *see* Side holes
Hyoscine-*N*-butylbromide use in stenting 84
Hyperplasia, intimal, transjugular intrahepatic portosystemic stent shunts and 204

Iatrogenic disease, postsurgical *see* Surgery
Ileocolic anastomosis, strictured 151–2
Imaging *see* Radiography
Inferior vena cava *see* Vena cava
Inflammatory disease, pancreatic gastrointestinal fistula related to 164, 165, 169
Injury, strictures caused by 53
Internal jugular vein, approaches via *see entries under* Transjugular
Intestine *see* Bowel
Intimal hyperplasia, transjugular intrahepatic portosystemic stent shunts and 204
Intrahepatic abscess *see* Abscess
Intrahepatic duct puncture for percutaneous biliary drainage 22–2
Intrahepatic strictures 53
Intubation *see also* Tube
  gastrointestinal strictures and 142–3, 146
  oesophageal strictures and 136, 137
Iridium therapy of malignant biliary strictures 30
Irradiation, internal, of malignant biliary strictures 30
Irrigation
  with biliary stones 9
  with gallbladder stones 73

J-guidewire for percutaneous biliary catheter introduction 22–5
Jugular vein, internal, approaches via *see entries under* Transjugular

Kensey–Nash lithotrite 78

Large bowel, embolotherapy 178 *see also* Colon; Rectum

Lithotripsy
  of biliary stones 8, 9–11, 64, 73–6
    choledochoscopy and 9–11
    extracorporeal 11, 64
    intracorporeal 8, 9, 9–11
    mechanical 8, 9, 10
  of gallbladder stones 68, 73–6
    extracorporeal 68
Lithotrite, rotary gall stone 78–9
Lobes of liver
  atrophy, percutaneous biliary drainage in presence of 20
  left, percutaneous biliary drainage approach via 21, 25
  right, percutaneous biliary drainage approach via 21, 22–5
Looping of drainage catheter outside liver 27–8
Lunderquist–Ring wire with pyloric strictures 147
Lymph nodes, para-aortic, biopsy 103
Lymphatics, opacification in percutaneous transhepatic cholangiography 16

Malignant tumours *see* Cancer
Manipulation catheters 24
Mesenteric artery embolization 178, 183, 187
Metallic stents/endoprostheses 34–5, 43–8, 61, 199–205, *see also specific types*
  biliary 34–5, 43–8, 48–9, 61
    advantages/indications 34, 50, 61
    complications specific to 41, 43–8, 89
    future developments 50
    replacement 89
    results 48–9
  vascular, in liver disease 190–205
    choice 205
    indications/contraindications 191, 196–7
    results 192, 203–4
    techniques 191–2, 197–203
    types 190, 205
Metastases in liver
  embolotherapy 176–7, 179–82, 184
  stenting 193
Methyl *tert*-butyl ether dissolution
  biliary stones 11
  gallbladder stones 77
Miller (double mushroom-tipped) stent 34, 36, 37
  insertion 36, 37
  occlusion, management 43, 46–8
Mono-octanoin infusion/perfusion
  biliary calculi 11, 64
  gallbladder stones 77
Mortality, transjugular intrahepatic portosystemic stent shunt-related 203–4
Mouth, oesophageal stricture dilation via 136, 137
Mushroom-tipped stent, double *see* Miller stent

Nasal route of oesophageal stricture dilation 136, 137
Nasogastric tubes, stricture dilation and
  gastrointestinal 142
  oesophageal 136, 137
Necrosis, pancreatic 114
Needles for percutaneous biopsy *see also* Biopsy
  aspiration *see* Aspiration biopsy
  choice 100–101
  directing 101–2
  track of, plugging, in liver biopsy with coagulation abnormalities 120–124
  visualization 105
Neff introducing system 23
Neoplasms, malignant *see* Cancer
Nitinol, stents and 50
Nose, oesophageal stricture dilation via 136, 137
Nutrition
  in gastrointestinal fistula management 160, 165
  with transjugular intrahepatic portosystemic stent shunts, postoperative 203

Obesity, morbid, strictures following surgery for 148–9
Obstruction, bile duct *see also* Occlusion; Strictures
  common 20
  endoprosthesis with 30–31, 32–51
  false localization of level of 17

hilar *see* Hilar duct
  malignant 18, 19, 30–31
  percutaneous biliary drainage 18, 19, 20
  percutaneous transhepatic cholangiography 14–17
Occlusion (referred to sometimes as 'obstruction')
  catheter 29
  portal vein, portosystemic stent shunt contraindicated in 197
  stent 43, 43–8, 50, 61
    management 43–8
Occlusion balloon, biliary stone manipulation with 8
Oesophagogram, post-stricture dilation 137
Oesophagus
  strictures *see* Strictures, oesophageal
  varices *see* Varices
Oral route of oesophageal stricture dilation 136, 137

Pain, poststenting, in right upper quadrant 87–8
Palmaz stent, vascular 190, 193, 199, 202–3, 205
Pancreas
  carcinoma, strictures associated with 82
  embolization *see* Embolization
  fluid collections, drainage 111–14
  inflammatory disease, gastrointestinal fistula related to 164, 165, 169
Pancreatico-enteric-cutaneous fistulae 164, 165
Para-aortic node biopsy 103
Parenteral nutrition with gastrointestinal fistulae 160, 165
Pelvic fluid collections, drainage 114–16
Percuflex, endoprostheses made of 34
Percutaneous biliary stone removal 11
Percutaneous biopsy *see* Biopsy
Percutaneous drainage *see* Catheters; Drainage
Percutaneous fine needle aspiration *see* Aspiration biopsy
Percutaneous gallbladder procedures 67–80
Percutaneous gastrointestinal fistula management 160–165, 165–9
Percutaneous stents 82, 89
  complications 87–8
  endoscopic and, combined 92–3
  endoscopic vs 82
  replacement 89
Percutaneous transhepatic cholangiography 14, 14–17
  incomplete 17
  strictures assessed via 53
Perforation *see* Puncture
Perfusion, stone/stone fragments cleared via *see* Chemical dissolution
Peristalsis, bowel, catheter and wire advancement through lumen and 145
Phlegmon, pancreatic 114
Plastic stents/endoprostheses 33–4, 49–50, *see also specific types*
  complications specific to 41, 43
    management 43
  future developments 50
  indications 50
  results 49–50
Portal vein occlusion, portosystemic stent shunt contraindicated in 197
Portography with transjugular intrahepatic portosystemic stent shunts 201, 203
Portosystemic stent shunt, transjugular intrahepatic (TIPSS) 190, 192–205
Postembolization syndrome 184
Postsurgical complications *see* Surgery
Precoccygeal/retro-anal approach to drainage of pelvic fluid collections 114
Prosthesis, endo- *see* Stents
Protein/polymer sheath, haemostatic 124
Pseudoaneurysm
  gastroduodenal artery (and branches) 178
  hepatic 178–9, 184
  pancreatic 182, 186
Pseudocyst, pancreatic, drainage 111–12, 113
Pseudocystogastrostomy 111–12, 113
Pulsed jet irrigation
  with biliary stones 9
  with gallbladder stones 73

Puncture/perforation
  bile duct, for percutaneous biliary drainage 22
  by endoprostheses (stents) 42, 87, 88
  of oesophagus
    stricture developing after 132, 133
    stricture dilation at site of 135
    stricture dilation resulting in 132, 133, 138
  as percutaneous biliary drainage complication 29
Pyloroplasty, vagotomy and, gastric stricture
    following 155
Pylorus
  catheter/wire advancement 143–4, 145, 146–8
  strictures 146–7, 148, 154, 157
Pyrexia, poststenting 87–8

Radiography/imaging *see also specific techniques*
  biliary decompression approach assessed via 19–20
  of biliary strictures 52–66, 83
    access via 54–60, 65
    assessment 19–20, 53–4, 63, 83
    biopsy guidance via 100–108
  intra-abdominal drainage procedures employing
    guidance by 108–16
Radiotherapy, internal, of malignant biliary strictures
    30
Rectum *see also* Bowel; Large bowel
  drainage of pelvic fluid collections via 114, 115
  embolotherapy 187
  intubation via, gastrointestinal strictures and 146
  strictures 149, 153, 158
Reflux strictures 129
Ring biliary drainage catheters 26–7
Rösch-modified Gianturco stent/endoprosthesis
  biliary 34, 35
    strictures treated with 35, 61
  vascular 190
    technique with 192
Roux loop, stricture access via superficially placed
    57–60, 65

Sclerosing cholangitis *see* Cholangitis
Sedation in stenting 84
Self-expandable stents *see* Expandable stents
Sepsis, transjugular intrahepatic portosystemic stent
    shunt contraindicated in 196
Shock-wave lithotripsy, extracorporeal
  of biliary stones 11, 64
  of gallbladder stones 68
Shunt, transjugular intrahepatic portosystemic stent
    (TIPSS) 190, 192–205
Side holes
  drainage catheter 27
  endoprosthesis 33
Sidewinder catheters in pancreatic embolization 183
Skin *see also entries under* Percutaneous
  catheter fixing to 28
  fistula between gastrointestinal tract and 160, 161,
    163, 164, 165
Small bowel (enteric) *see also* Duodenum; Ileocolic
    anastomosis
  embolotherapy 178
  fistulae involving 160, 161, 163, 164, 165
  strictures involving 151–2, 157–8
Sonography *see* Ultrasound
Sphincterotomy, endoscopic
  in ampullary carcinoma 91
  biliary stone removal via 11
  for stenting, complications associated with 87, 88, 91
Steerable catheters, biliary stone extraction
    procedures employing 4, 5
Stents (endoprostheses), biliary 28–9, 30–31, 32–51,
    82–95
  complications and their management 41–8, 61
  design 33–4
  drainage catheters vs 28–9, 32
  endoscopically-inserted *see* Endoscopic stents
  functional assessment 85–6
  future development 50
  percutaneously-inserted *see* Percutaneous stents
  replacement/removal 43, 88, 88–9
  results 48–50, 90–92
  stricture treatment with 60–61, 82–93
    postdilational 60–61, 85

technique of insertion 28, 36–41, 85
Stents (endoprostheses), vascular, in liver disease
    189–205
  indications/contraindications 191, 196–7
  results 192, 203–5
  techniques 191–2, 197–203
  types
    choice 205
    description 190
Sterility, ultrasound transducer 105
Stomach *see entries under* Gastr-
Stones (calculi)
  biliary 3–12
    impacted 7–8
    intrahepatic 8
    management (removal/destruction) 3–12, 63–4
    stricture-associated 63–4
  fragmentation *see* Lithotripsy
  gallbladder 71–9
    impacted 73, 76
    management (removal/destruction) 68, 71–9
Strictures
  biliary *see* Strictures, biliary
  gastrointestinal 141–58
    lower 149, 157–8
    upper 146–9, 154–7
  oesophageal 128–39
    aetiology 129–33
    management 133–9
Strictures, biliary 52–66, 81–96 *see also* Obstruction
  access, radiological 55–60, 65
  aetiology/epidemiology/location/presentation 53,
    82, 83
  anaesthesia 54
  assessment 19–20, 53–4, 83
    of extent 19–20
    radiological 19–20, 53–4, 63, 83
  benign 52–66, 93–4
  calculi associated with, management 63–4
  contraindications to intervention 54
  drainage catheter not advancing through 28
  drainage catheter positioning regarding 27
  hilar *see* Hilar duct
  intra- vs extrahepatic 53
  low 90–91, 92
  malignant 82–93
    internal irradiation 30
  patient preparation 54
  treatment 60–65, 83–96
    complications 65, 87–8
    follow-up 61–3
    results 65
Subcostal approach
  percutaneous cholecystolithotomy 75
  percutaneous cholecystostomy in cholecystitis
    68–70
Subphrenic abscess percutaneous biliary drainage
    complicated by 29
Suppurative cholangitis, percutaneous biliary
    drainage in 24
Surgery
  cholecystectomy via 68
  strictures associated with
    biliary 53, 94
    gastrointestinal 147–58 *passim*
    oesophageal 132, 133
  strictures managed via
    biliary, radiological intervention in relation to
      65
    oesophageal 133
Suture anchors in percutaneous cholecystolithotomy
    72

T-tube, biliary stone removal employing 4–8, 9,
    9–11, 11
Tandem method of biopsy needle placement 101–2
Telangiectasia
  gastrointestinal 178, 184
  hepatic 185
Tracker 18 system in embolotherapy
  of liver 179
  of pancreas 183
Transcaval tumour biopsy 119–20

Transducer, ultrasound, in percutaneous needle
    biopsy guidance 104–5
  choice 104–5
  sterility 105
Transgastric approach to pancreatic pseudocyst
    drainage 111–12, 113
Transgluteal approach to drainage of pelvic fluid
    collections 116
Transhepatic access/approach
  percutaneous cholecystolithotomy 74
  percutaneous cholecystostomy in cholecystitis
    68–70
  strictures 55
Transjugular intrahepatic portosystemic stent shunt
    (TIPSS) 190, 192–205
Transjugular liver biopsy 119–20
Transnasal route of oesophageal stricture dilation
    136, 137
Transoral route of oesophageal stricture dilation 136,
    137
Transplant candidates, liver, portosystemic stent
    shunt for prospective 196
Transrectal approach to drainage of pelvic fluid
    collections 114, 115
Transvaginal approach to drainage of pelvic fluid
    collections 115–17
Traumatic strictures 53
Trucut needle in plugged liver biopsy 120
Tube(s) *see also* Intubation
  biliary stone removal employing 4–8, 9, 9–11, 11
  biliary stricture access via operatively placed 56
  drainage, gastrointestinal fistula-associated
    abscesses and 162
  gastrostomy *see* Gastrostomy tube
  nasogastric, oesophageal stricture dilation and 136,
    137
Tumours
  benign, of liver, embolotherapy 177, 179
  malignant *see* Cancer

U-tube, stricture access via operatively placed 56
Ultrasound/sonography
  biopsy guided by, percutaneous needle 104–5
  cholangiography guided by, percutaneous
    transhepatic 16, 53
  cholangiography preceded by, percutaneous
    transhepatic 14
  drainage procedures employing, percutaneous
    109–10
    biliary 19–20
  lithotripsy by, with gallstones 76
  portosystemic stent shunt guidance by,
    transjugular intrahepatic 197–8, 198, 200
  stent function assessed via 85, 87
  stricture access employing 54
  stricture assessed via 54, 63

Vagina, drainage of pelvic fluid collections via 114–16
Vagotomy and pyloroplasty, gastric stricture
    following 155
Varices, gastro-oesophageal 199, 202–3
  bleeding 196, 199
  embolization 199, 202–3
  endoscopic assessment 203
Vascular stents *see* Stents
Vasovagal reactions, oesophageal stricture dilation
    and 138
Vater, ampulla of, carcinoma, management 91
Vena cava, inferior
  stenting 190–192
  tumour biopsy via 119–20

Wall, bowel, catheter and wire advancement through
    lumen and distensibility of 145
Wallstent 35, 37–41, 43–8, 61
  biliary 35, 37–41, 43–8, 61, 89
    insertion 37–41
    occlusion, management 43–8, 89
    results 48–9
  vascular 190, 205
Wholey wire in strictures
  of oesophagus 136
  of pylorus 147